Today's Facts

Today's Facts

Understanding the Current Evolution of Information

James W. Cortada

ROWMAN & LITTLEFIELD
Lanham • Boulder • New York • London

Published by Rowman & Littlefield
An imprint of The Rowman & Littlefield Publishing Group, Inc.
4501 Forbes Boulevard, Suite 200, Lanham, Maryland 20706
www.rowman.com

86-90 Paul Street, London EC2A 4NE

Copyright © 2025 by The Rowman & Littlefield Publishing Group, Inc.

All rights reserved. No part of this book may be reproduced in any form or by any electronic or mechanical means, including information storage and retrieval systems, without written permission from the publisher, except by a reviewer who may quote passages in a review.

British Library Cataloguing in Publication Information Available

Library of Congress Cataloging-in-Publication Data available

ISBN 9798881804732 (cloth) | ISBN 9798881804749 (ebook)

∞™ The paper used in this publication meets the minimum requirements of American National Standard for Information Sciences—Permanence of Paper for Printed Library Materials, ANSI/NISO Z39.48-1992.

Contents

Preface		vii
1	Learning from the History of Information	1
2	How Information and Computers Changed Work	35
3	Do We Live in an Information Age?	63
4	The Emergence of Big Data	85
5	How Factual Is Information and Why Should We Care?	111
6	A Way to Look at Information Today	135
7	The Special Issue of Artificial Intelligence	167
Notes		179
For Further Reading		221
Index		233
About the Author		243

Preface

Our world is so full of information that we have become its prisoner. We cannot seem to do anything without consulting collections of organized information curated by experts, pundits, practitioners of the high art of tossing out opinions on all manner of topics regardless of whether they know what they are talking about, and software. People rightfully complain that they are drowning in information—they have been doing this for centuries, of course—but nonetheless they are right to grouse, because we do have so much of it. Their information is fragmented into "fields," "subjects," "disciplines," or "specializations," meaning that we are finding it difficult to holistically view a subject, to take into account the "big picture," yet we knew a thousand years ago and even this morning that context is at least as valuable when making a decision or forming a point-of-view than any specific narrow set of facts. But all this information, both good and bad, accurate and inaccurate, wonderful or malicious, affects the quality and serenity of our lives. It seems nowhere is this more so than in societies that are generating the most facts, data, information and seemingly less insight and wisdom than we desire. Understanding the contradictory phenomenon of rising levels of information—think tsunami—and its ever-more fragmenting features, therefore, becomes a central aspect of modern life essential to appreciate.

It has been my purpose to facilitate that understanding for a half century. The book you are reading continues my exploration of the modern form and uses of information, focusing largely on the American experience, as it is the one I have studied the most. However, much of what is discussed in this book is applicable in most societies around the world. Today pundits and wizards warn that artificial intelligence (AI)—generative versions now—could become more intelligent than humans and take over the world. We mere mortals are horrified at the thought, as we should be, because we like running the tables in our casino of life. The Bible and other religious tracts tell us that Earth is our paradise to use, so how dare an algorithm and its metallic carcass, the computer, take over our monopoly? Are we fighting a losing battle against these cyber rivals that really do not care about global warming and changing environments since they can stand more heat and cold and less clean air than humans? We think not. We use information to fight back and to nurture and exploit our paradise. But we must understand what is happening if humans are to continue thriving. Information is ultimately our best weapon—think of it as a tool—for doing that. This book is a small contribution to our arsenal. Now let's get to it.

Information is as visible a part of modern society as is the small computer you hold in your hand, but that you still call a smartphone. How quaint, even charming. Yet, it is a remarkable device that can bring all the world's digitized information to you anywhere as image, text, or voice. It can send messages, but that you can use like a telephone (another old-fashioned term). This little device is emblematic of almost everything we know about information. With billions of copies carried around almost with the same diligence as one does not forget to wear clothes before stepping out, access to information has become almost an extension of our bodily functions; certainly, these little machines are almost appendages. They continue to evolve, acquire additional functions, and become lighter and thinner. These are the current culmination of some two hundred years of humankind's attempts to collect and use information. Electricity, then telegraphy, telephony, and radio—all inventions of the nineteenth century—were the great-grandparents of this device. In the next century came computers, more telephony, software, and databases—the grandparents of this same device. Improvements in material sciences and computer chips and batteries made it possible to stuff into one of these the equivalent contents of a major university library, access to trillions of amounts of information beyond the holdings of the U.S. Library of Congress. And you do not need a PhD in computer science to operate it. You almost needed that for its grandparents' generation of mainframes of the 1940s and 1950s.

Portability of the smartphone further caused people all over the world to become ever increasingly reliant on more bodies of organized information in the past two decades than probably was the case in all of human history. That dependency has grown over many centuries, however, and no more rapidly than in the past 175 years. The smartphone is a natural and logical extension of that process facilitated by the development of the physical artifact itself. It is often seen as a metaphor for our times, while becoming so ubiquitous that we hardly acknowledge its presence. How dependency on information, and even the very nature of how facts came about, has been the subject of investigation by every discipline. Historians are beginning to piece together how that fundamental change in human dependency came to be. This book summarizes strands of those investigations.

This is the third and final volume in a broad study about the role of information since the early nineteenth century. The first volume, *All the Facts*, established that organized bodies of information had existed and had been used extensively in the United States since the seventeenth century, but with a massive expansion in its amount and diversity after the mid-nineteenth century. Its message was essentially "everyone" was using facts, and most of it came in published form. That study extended to the twenty-first century with all its computers and Internet. The second book, *The Birth of Modern Facts*, observed that the introduction of electricity with which to move information about and the development of new professions and bodies of facts to support these in the nineteenth century helped to transform information, not simply added to by more users. This occurred at the same time as people were becoming more reliant on information, in part because of so many new types that were becoming available, an expansion that continued to the present.

In this third volume, I summarize how information changed, arguing that the development of new professions stimulated more transformations in support of those jobs and interests than the arrival of the also very important technologies embodied in electricity, telecommunications, and later, computers and the Internet. That argument brings balance back to the history of how information came to be so important, as it required professions, new informational discoveries, and computational technologies. It was the combination of what was explored in both prior volumes that began to give us a better sense of today's role of information. It was that combination, too, that is affecting the current evolution of information, most obviously in such areas as genetics, DNA, quantum physics, and even life itself. You will learn about what biologists and developers of AI will be doing over the next decade, but also that their successes built on the patterns of behavior identified in the two earlier volumes.

Here, I pull together key findings and consequences from those two studies and integrate the work I did for a prior trilogy, *The Digital Hand*, which looked at how computing and information transformed the work of over a dozen large American industries after World War II. Woven into this current volume will be findings and their significance drawn from other studies on such matters as information ecosystems, infrastructures, and fake news—much about current events. While for two centuries fragmentation of types of information took place, with Big Data and AI we are possibly witnessing the reintegration of bodies of data (facts), a dream librarians cherished for centuries. As a result, we may have stumbled across a new consequence of the activities involving information of the past two centuries; so it cannot be ignored. This book essentially answers the very rude but useful question

Americans like to pose: "So What?" It is an attempt to provide an answer through a deeper understanding of information's history that cannot be gleaned from the many thousands of short books and articles, often written by commentators responding only to immediate circumstances. I do that by defining at a high level various patterns in the evolution of information. The table of contents is straightforward, and each chapter early on explains its role and how we proceed through the text. You can read chapters in any order, but I encourage you to look at the first and last ones, at least.

This book has been a journey of well over four decades as I explored the histories of computers and their uses and users, collaborated with others on studying the history of information and computing, and now it is the subject of these three books. I looked at computing and information through the lens of a user and corporate employee inside the computer industry, reporting results in business publications, which allowed me to be in the arena that was making history, while observing it unfold through the eyes of the historian.

Finally, let us acknowledge an obvious reality: that scholars from across many disciplines are trying to codify, or at least connect, various pieces of the story of information's nature and role. If you sent the title of this book to five recognized experts from five different disciplines and asked each of them to come up with a table of contents for such a volume, you would receive five very different responses, and all of them would be appropriate. I know because I had five different conversations with experts while thinking through what this book should cover. What that exercise tells us is much is still in flux. But we do know that historical perspectives provide context for understanding current activities, and since I apply a historical perspective to the conversation in this book, it helps to further our appreciation of the role of information in today's society.

In each of the earlier publications, I recognized and thanked the many people who helped me on my journey. But there are two people who need to be recognized here because they pointed the way, helped, but also got out of the way as I went on my trip. The late Al Chandler, business historian and friend, and I had a number of conversations about how to look at information through business and economic disciplines, because I had only flirted with concepts involving knowledge workers. He also agreed to collaborate on a project to get me going. Five books later, Bill Aspray pulled me aside to suggest I write a book about the role of information in corporations and to deemphasize my interest in the uses of computers as a technology topic. His suggestion caused me to shrink the role of and my focus on computers *per se* and to expand my examination of the role of data, facts, information, knowledge, and wisdom in the way entire societies worked. That is how I arrived to this page you are reading. I grew in my understanding of modern societies; I hope you do as well.

Jon Sisk, editor of this book and its predecessor, *Birth of Modern Facts*, continued to support my work; there is nothing like having an experienced editor on your team! The production team at Rowman & Littlefield/Bloomsbury Publishing once again turned a sprawling manuscript into the book you are reading, so many thanks to this crew. The views and weaknesses the reader encounters are of my own doing. The views expressed in this book do not necessarily reflect those of the publisher or of my colleagues at the Charles Babbage Institute at the University of Minnesota—Twin Cities. My journey is not done, so I encourage you to share your thoughts and coach me on what to explore. I can be reached at jcortada@umn.edu. Don't be shy.

—James W. Cortada

Chapter 1

Learning from the History of Information

Juger un homme par ses questions plutôt que par ses réponses (Judge a man by his questions rather than his answers).

—Voltaire (a.k.a. François-Marie Arouet)

Voltaire (1694–1778) admonished his peers, intimidated subsequent generations of scholars, and, by inference, too, students of the concept of information by arguing that asking questions was more important than providing answers. While his epigraph above advised on how to judge a person, his underlying point was more about inquiring and consequently that answers to questions change. He would have understood what scholars in many disciplines have been experiencing for over two centuries in their exploration and development of information. That he died over half a century before the start of the Second Industrial Revolution, and the resurgence in the quest to organize and create new information, is a testimony that before mid-century thoughtful commentators were already conditioning future generations to pick up such tasks and carry them forward. But why quote Voltaire rather than other French thinkers, such as Jean-Jacques Rousseau (1712–1778), or Scottish enlightenment thinkers, such as Adam Smith (1723–1790) or David Hume (1711–1776)? He may have better understood the problem we face in this book: the challenge of dealing with all manner of information, because he wrote literature, philosophy, essays about history and political affairs, and explored what eventually came to be known as the hard sciences. If blogs had existed in his day, undoubtedly he would have commented on all manner of contemporary issues. As it was, he wrote some 20,000 letters and a combination of roughly 2,000 books and pamphlets.[1] The others, however, understood that the quantity of information on many topics was increasing beyond what one individual could absorb, but Voltaire tried.

The Frenchman's eighteenth-century concerns have become worse. Since his day, people have complained that there is too much information; they were right. In his day, scientific knowledge (information) increased at an estimated 2–3 percent between then and the 1930s and by 8–9 percent since World War II. One study supporting those growth rates suggests the magnitude of the volume just for the hard sciences, analyzing in excess of 755 million references in 38 million publications published between 1980 and 2012.[2] Another examined 1.8 billion publications across 241 subjects—that is a massive amount of information housed in these publications.[3] There is now so much information that the rate at which users of today's information can absorb it is slowing. Symptoms are everywhere: Nobel Prize winners on average taking 10 years longer to make their contributions than before World War II, more years spent in education for someone to get their intellectual arms around a field, further specialization, and the inability to identify new and important ideas. So scholars, in particular, keep relying on older, more familiar bodies of knowledge, which means many new findings may possibly never get an airing. Quality of life, dealing with environmental problems, overpowering medical threats, and so forth are at risk, because of too much information, or what one student of the problem called the growing "burden of knowledge."[4] In other words, information is a problem and therefore we need to understand it better than we do. Along the way, users (a.k.a. "experts") are being accused of mismanaging information as the public wades into the waters of data, context, and misinformation. An essay in the *New York Times* captured the essence of the discord with the title, "We Need to Save Expertise From the Experts."[5] In this book, I discuss what has been happening with information in the past several decades when increasing numbers of people have been using organized information in greater quantities, in newer ways, relying on information handling technologies, and for an expanding set of purposes.[6]

Of course, the vast increased use of organized information by the public at large hides another truth that has barely been recognized. Peter Burke, an expert on the role of ignorance, explains: "Although humanity as a whole knows more than ever before, most individuals know little more than their ancestors did."[7] He further observes that "as information continues to pile up, there is more and more for each one of us not to know."[8] While one could challenge that level of ignorance, it is possible to slightly challenge—qualify—his observation by pointing out that fragmentation of large bodies of information, a result of deeply informed specialties (disciplines), facilitated the limitations of facts the public at large might access.

Today, most writers produce less than did Voltaire, and their materials are more narrowly focused. That single change in behavior—managing scope and specializing in narrower topics—is a direct byproduct of asking questions more often than just providing answers. That altered behavior, more than any other feature of modern human activity, accounts largely for the changes in the variety and volume of information that humans alone have created since Voltaire's time. If one is to believe biologists, humans are not the only living beings creating diverse bodies of information.[9] They are sufficiently persuasive to require the exclamation that this book is focused on human relations with information. I concentrate on the anthropomorphic. It would not have been necessary to provide such a qualifier in the twentieth century, but to ignore the wider subject of information is to deny the potential breadth of the topic that scholars are increasingly facing in the twenty-first century. One can surmise that Voltaire would have loved meeting mathematician and electrical engineer Claude E. Shannon (1916–2001) or biologist James Watson (b. 1928) to discuss electronic communications or DNA, both of whom profoundly influenced our understanding of information in the twentieth century. So, we will not discuss how trees communicate with each other, or how birds or squirrels chatter.

Historians and other students of information (facts, data, knowledge, even insights and wisdom) have long studied the subject, but never so methodically as in the past seven decades. They cataloged bodies of facts and subjects, identified features of these, and most recently engaged in lengthy debates about the definition of information.[10] In the process, historians identified the amount and extent of use of organized bodies of facts, proceeding far beyond long-standing practices of discussing histories of books, publishing, and libraries. Their obvious conclusion that there was a great deal created, far more so since the mid-nineteenth century, has been augmented by more specific findings about the types created and used, and by whom.[11] This chapter summarizes some of their findings. Yet, the topic remains contentious, with even the definition of information subject to ongoing debate.[12]

We proceed by discussing unsettled definitions of information, how much exists, and more importantly, how it has been changing over the past century and a half. Because of insights emerging from the work of biologists, their work is added to the discourse, followed by an assessment of the effects of technologies on information. I conclude with a description of information based on the combination of historians' findings and how scientists are shaping the same subject. Why? As our understanding of a subject evolves, it is useful to summarize where we are to assist in the discovery of further findings, or to influence their evolution. Historical experience suggests that periodic updating of perspectives based on consolidating new facts and thinking about a topic assists in the further advancement of knowledge and the application of new information. While this chapter may appear arrogantly broad, it carefully links to my research findings explained in greater detail in *All the Facts*, and in *Birth of Modern Facts*, and to that of other researchers.[13]

To attempt clarity, each section of this chapter is entitled with a finding that can assist others in their future work on the subject.

THAT DEFINITIONS ARE STILL UNSETTLED

My name is Jim—that is an uncontested fact. Tuesday follows Monday, while water freezes at thirty-two degrees Fahrenheit. We think we know what a fact is, and in recent decades, facts and information have been considered synonymous terms; for convenience, I use them interchangeably in this book. Meanwhile, variation in what is information remains subject to interpretation by the public at large, and certainly too, by scholars studying the subject, such as the definition of knowledge versus, say, wisdom. No matter, much of what people encounter are facts: my name is still Jim, a particular individual at work is still a terrible manager, and everyone there knows that is a fact. Or is it? There are pieces of information that are accepted as absolute, such as your name, but others are subject to interpretation, such as the managerial skills of a colleague. Further to the edge, there are "facts" subject either to interpretation or which are categorically or inferentially believed to be wrong, fake, or misleading. Philosophers have told us for millennia that how we think about "truth" (their word for facts)—logical thinking augmented by storytelling, or what some today call "narrative intelligence"—is a significant part of what we think of as information.[14] So, in fact, the definition of information is subject to varying interpretations.

Historians of technology and the hard sciences, in particular, have encountered this problem of definitions because the variety of possible interpretations of what one means has increased over the past two centuries, due to changing collections of data and interpretations. They remain concerned even after all this time. Historian of technologies, Rosalind Williams, raised the issue as late as 2021, echoing admonitions of earlier scholars: "be prepared to explain what you mean by it," and "to defend its use, and to ask yourself if there are more precise or accurate words to think with."[15] She drew her readers' attention to the need for specificity in the context of discussing what crisis means in the history of technology, while broadening her admonition to the wider issue of definitions involving facts, concepts, procedures, and context in broader terms. I have concentrated much of my research about information on this same point—the need for specificity in what we mean— and it is difficult to achieve.[16]

Philip Tetlow, long steeped in the technology of the Internet, made a similar, related point: that context matters. Thinking of the differences between data ("On its own, data is meaningless") and its value in context (e.g., a number representing someone's birthdate) are essential.[17] Biologists have explained, for instance, what Tetlow describes as the "intracellular genetic processes" to "provide a self-contained mechanism for making sense of the various elements needed to provide the data and structural elements" required to function, served up by chemical concentrations of descriptive and positional proteins. These are offered up to provide the wherewithal: "appropriate developmental context for the right cells to form in the right place at the right time."[18] Most useful from Tetlow is his observation that "this inbuilt choreographed cascade of combining complexity results in the elegant and beautiful system we indifferently refer to as life."[19] He observes, as part of the Internet, I too in how information ecosystems function, that such behavior exists outside of living matter as well. If all these ideas about biology, the Internet, and how humans think and act seems like a mouthful, well, one would be right to think so. It is why we encounter such issues throughout this book. But as a reminder, since we need to strive for simplicity, psychologist and information expert Andrew Dillon advocates the view "of our mental life as a form of information processing, with our minds serving as both a filter and storer" of all manner of information, including from our senses, light, sound, chemical data and, I add, from information.[20]

The unsettled nature of definitions is exacerbated by people who are extensive users of organized information, such as doctors, teachers, accountants, professors, media experts, law enforcement, public officials, and scientists, among others. To apply a popular phrase used in business circles, "thought leaders" present points of view in which they organize data, facts, or information to exclaim yet another fact. This is done, for example, when a lawyer argues that his client did not commit a crime, or by a liberal politician declaring that their conservative rivals are destroying democracy, or by a newspaper columnist accusing someone of promulgating fake facts or, to use an old-fashioned term, of lying.

However, historians are increasingly discovering that users and keepers of organized bodies of information have engaged in debates about what constitutes information, such as what appears in books and computerized datasets, while in the process of creating collections of facts needed to do their work or what they uncovered in the process of working.[21] Academics and others who deal with published bodies of information or in the sciences, more than other groups in society, have shaped definitions and features of information over the past two centuries. These include professors, others in library science, the hard sciences, increasingly in media studies, and in the humanities and social sciences.[22] They have not been shy about promoting discipline-specific definitions. Engineers, computer scientists, and others working in such industries as chemistry, pharmaceuticals, steel and material culture, and manufacturing, added descriptions of information. All have been affected by the fact that the information upon which they rely kept changing. "Progress" in medicine is an obvious example, but, too, in such areas as flight, microelectronics, telecommunications, and so forth. One of the most influential thinkers about science in the twentieth century, Thomas Kuhn (1922–1996), argued that scientists accumulated new information and theories to an extent that eventually accepted existing understandings no longer aligned with new facts and emerging theories, so old facts are dropped and new ones embraced.[23]

Ongoing development of new types of information still underway sped up in the past several decades of the twentieth century, due increasingly to the work of scientists and biologists, for example. That activity shows no signs of slowing. But it is a process with deep roots. There were dual central influences on the nature of information in the past two centuries. One was the introduction of electrified information, beginning with the development of telegraphy in the 1830s, which extended to telephony, radio, television, computing, and the integration of computing and telecommunications. These developments led to new notions of what constituted information, extending the concept, say from words (e.g., Jim is my name), to include electrical impulses (positive and negative ones, telegraphy's "dots" and "dashes," and computing's "1s" and "0s"), or to what Claude E. Shannon referred to as "noises" and "signals" in the transmission of electrified information.

The second influence was the emergence of professions and academic disciplines. Examples of professions include accountants and managers, of the second physics and psychology. In combination, all shaped information. All professions also expended enormous energy in organizing their collections of information into typologies relevant to users, most specifically within their own disciplines.

Several reasons account for the energy spent in organizing information. The *variety* of new information expanded enormously, so getting to what someone wanted proved to be an increasingly growing challenge. Every profession and discipline fought the problem. These still do today, with much of the shoveling against the tide of more information on the beaches of websites and social media. *Complexity* of new information, too, added to the problem. Recall our brief discussion of biology served up by Tetlow, a computer engineer. He pointed out that in the case of the Web it is being assaulted by many structures, in his opinion with "too much."[24] His complaint—one resonating increasingly with those who ponder the massive participation of the public at large in social media, for example—is that "people just keep pretending that they can make complex things deeply hierarchical, categorizable, and sequential when they can't accomplish this directly."[25] He has a point. Over time, information became entwined with increasing amounts of other information. That may explain why when one Googles a phrase, they are presented with over a million "hits." The result of overlapping hierarchies and topologies, and so forth, leads to one's inability to not see the forest for the trees.[26]

Overcoming this problem is fundamental to many of the new generative artificial intelligence (AI) tools introduced to the public in the early 2020s.

In the process of understanding information as an evolving long-standing historical process, scholars have come to appreciate the role of rules in organizing human behavior and physical realities through natural laws and laws of nature.[27] This recognition extends even to the invention of our modern-day definitions of facts (taxonomies) in the Middle Ages,[28] and the grand progenitor of modern information development one refers to as the "scientific method" of discovering and using data, facts, and information. To put the latter into a sound byte, scientists define the scientific method as "a systematic way to build knowledge with foresight at its core."[29] The same scientists just quoted remind us, however, "much of human power derives not from our understanding foresight per se, but from our understanding of its strengths and weaknesses."[30] Three steps are involved in the modern understanding of a phenomenon applying scientific methods: collection of data through observations and experimentation, developing potential explanations (understanding) of what these data mean, then developing hypotheses drawn from such explanations that can then be tested. That three-step process goes far in helping to explain modern definitions of information. At least for historians, this process helps to explain why we have scientific journals (and their role) and why and how we have developed methods for categorizing and accessing information quickly, most notably by way of a nearly 1,000-year process of developing and applying the humble index.[31]

In the twenty-first century, biologists refined their understanding of what living matter does and how it uses information. Their work holds out the possibility of affecting profoundly our understanding of what information is and how it functions, to the extent that the work of Shannon and others did by the mid-twentieth century. DNA molecules house genetic information; how that occurs will not detain us here. It is important to acknowledge that information contained in DNA is used to guide the activities of proteins and other living matter, a process still being studied or, to use our focal point, to be defined.[32] What is increasingly recognized, however, is that DNA can be linked to notions of information storage. Even computer scientists are interested because of the amount that can be stored in a tiny space by living matter. Biologists have long known that how "Mother Nature" constructs matter, life forms, and behaviors favors the simplest approach. Scientists have come to recognize that the simplest explanation for a phenomenon, too, often provides a better explanation of how something works. One computer technologist answers the obvious question "So What": "I have become increasingly amazed . . . that we still commonly construct computer systems, especially complex computer systems, using methods and patterns different from those chosen by nature in the fashioning of its own solutions."[33] Taken to its logical conclusion, perhaps computer scientists and engineers should treat the Internet as a living being and thus apply nature's way of evolving and responding to physical realities.[34] Biologists are increasingly concluding that information is dynamic, a force of its own, and organized as networks, rather than perching in hierarchies. For them, information is not just a noun but also a verb.

Historians still have insufficiently studied the role of different groups influencing the definition of information, such as artisans, blue-collar workers in the "trades," and craftsmen. Historians have paid more attention to academics, engineers, and so forth, less so to others who developed new bodies of information. Those working outside of academia and scholarship addressed problems that came up in their work but did not, like academics, rush to scholarly journals to opine on the nature of information or formulate definitions and typologies. So, workers and artists, for instance, remained beyond the pale of many discussions about definitions. They, however, had points of view on the matter. They viewed information as practical, almost as objects, as tools.[35] Facts and information were synonymous, made valuable if wrapped in precise instructions on how to apply these.[36] In his study of the father of modern economics, Adam Smith, economist Benjamin M. Friedman reminded us not to leave those classes of workers out of the discussions, because as they specialized they became more productive economically.[37] Also keep in mind, people think of information as objects and tools, to facilitate the mundane acts of living, from fixing a lawn mower to cooking a new dish, for example.

Specialization led to more information on unique varieties. An information technology historian, Nathan L. Ensmenger, made a similar point regarding computer programmers.[38] Programmers were not PhDs in computer science or engineering. Some were mathematicians, others chess players and radio ham operators, even seamen in a navy who learned basic electronics. Many early and even contemporary programmers trained without the benefit of any university coursework, rather through on-the-job experiences, or by corporate trainers/practitioners. One could convincingly argue that they developed their craft, and in some instances early databases, the way blacksmiths, farmers, and other craftsmen created theirs. Earlier programmers prided themselves on the elegance of their software, much as someone might in developing a beautifully crafted silver teapot or a philosopher an elegant argument. It was art. Is art a form of information? Certainly, the methods and tools used to create it were considered crucial components of an artisan's "tool kit." The point is, society continues to develop new classes of craftsmen and artisans who in turn create their own bodies of information, skills, and tools, and so treat these nearly as objects to be applied in solving problems and doing work.

This process existed long before today as, for example, when Europeans developed production processes in earlier centuries, even more in industrializing societies in the nineteenth, and massively so in the twentieth. Expertise, knowledge about how to apply new techniques, and codification of practices ensured that all manner of workers and scholars engaged in the codification of newfound knowledge applied to all manner of problems and issues.[39] Each brought to the table their vocabulary and notions of what constituted information.

So, the problem of evolving definitions is large enough that most scholars writing about the history, or role, of information still feel compelled to devote attention to offering their own. We will not wade through that discussion, as it could consume the rest of this book. While I have done the same in both predecessor volumes to this one, one can see the variation still emerging in all subsequent chapters.[40] The evolution of definitions shows no signs of settling. Rather, so much new and types of information are being created that attempts to solidify a definition seem premature, or, borrowing from psychologists and other scholars who today are fashionably called "brain scientists," it may be that the public and essentially all academic disciplines are aware that information plays a central role in our society. Historians have argued that information has long played a crucial role, and in more recent years in formats that we are accustomed to today, such as in books and magazines. Librarians and academic disciplines organized these various collections of information.[41]

I take it that there is a scientific basis, too, for such assumptions about the centrality of information—an issue addressed frequently in this book. But for a specific issue, we can call up "frequency illusions," a behavior experienced possibly by everyone at one time or another. You buy this year's edition of a Nissan Altima car, drive it around your community for a few days, and notice that many people have acquired the same model, even in the same color. Amazing! Another: you are seven months pregnant and notice that there are many pregnant women all over the place in shops and supermarkets you frequent, dropping children off at school, working in offices, well, just everywhere. What a coincidence; it must be the Year of the Baby. Neither is true. People are not rushing to Nissan dealerships to acquire this year's model any more than in earlier ones, and the U.S. Bureau of the Census has been reporting fewer births over a number of years. Impossible, you say? Known as the "frequency illusion" or as the Baader-Meinhof phenomenon, your brain plays a mind game psychologists elegantly refer to as a "cognitive bias." This is where your brain has a tendency to notice a particular phenomenon, such as more Nissan cars or pregnant women. It occurs after noticing something for the first time, such as a specific model of an automobile because you just bought one, causing you to see other copies more frequently when this happens, creating the illusion that something is appearing more frequently when it is not. After owning the new car for a few weeks, you will not notice as frequently others exactly the same; other pregnant women are subsequently hardly noticed too. We do not need to linger on the history of this behavior, its naming only occurring in the 1990s, so newly identified.[42] There are many types of cognitive biases, often described also as *inclinations* or *tendencies*. We care about such brainy matters because what constitutes facts—often also called truths or beliefs—helps shape what is information, such as truth and falsehoods, or misleading (e.g., the supposition that many people just bought a Nissan car). All affect what one perceives as a fact, hence its definition or subject.[43]

But to pull together these various strands of discussion on definitions, there are several realities being increasingly recognized. For one, information and its creation is a social process. Once a community accepts a fact, it becomes actionable and usable because others validate it. This happens, for example, when a scientist uncovers a finding, writes a paper that is peer reviewed, next published in an appropriately credible journal, and then accepted by a wider audience as true. The behavior of this community—its norms and values—becomes the basis for shaping the definition and value of facts. As one student of the process explained, scientific knowledge is "not merely as the product of individual or even group effort but as an emergent property of interactions across a social network," echoing what we discuss later about the role of information ecosystems.[44] Acknowledging the role of group interactions complicates the definition of information. Its definition is still under development at a time when grand theories of anything are in short supply, but highly desirable, because of our next finding.

THAT A GREAT DEAL OF INFORMATION EXISTS

The amount of organized information, that is to say, facts presented in publications, databases, other digital files, in libraries and bookstores, in magazines and other published materials, including manuals and instructions, textbooks, PowerPoint presentations, and so forth, has increased enormously in the past half millennium, and even more annually over the past several hundred years. Historians have better documented this fact than any other feature of information, especially regarding books and scholarly journals. Perhaps they understand the quantity involved because it was, to be candid, relatively easy to track. During the past two centuries alone in the United States, but similarly in Western Europe, the number of books, for instance, increased both in new titles and in copies printed.[45] Specialization of work and jobs did, too, sharply after 1850, and information needed and desired by people (even children) to go about their activities. In the United States, rates of literacy had been high since the 1600s in comparison to those in European nations. With the exception of indigenous Indians and African Americans, for Caucasians, these were over 65 percent in the nineteenth century and exceeded 90 percent during the next century. The act of teaching literacy normally involved their reading nonfiction (information), others on religion (usually the Bible), and books, newspapers and increasingly after the Civil War in magazines.[46]

All of this activity was compounded by the spread of formal education and the requirement to attend classes for more years, increasing incrementally in number over time. In the early 1800s in the United States and in some parts of Western Europe, a child might only attend school for several years, but by World War II, the majority were expected to complete twelve years through high school. Today, a third of the U.S. population aged twenty-five or older has completed college-level education, 90 percent have completed high school.[47] All of this activity increased, too, by the growth in population. In 1840—the decade most historians accept as the launch of the Second Industrial Revolution—the American population stood at 17 million, in 2020, it exceeded 331.5 million.[48]

The shift from an agricultural to a largely industrial economy, which so characterized the evolution of nineteenth and twentieth-century work, was in large part driven by the development of new information that made possible novel services and products. Electricity brought lights at night, telegraphy, telephony, and more practical engines. Development of steel made possible stronger building materials leading to skyscrapers, for example, but too, many other industrial products (e.g., new types of bridges) and home appliances. One can track innovations by counting patents. In 1790, only three were issued in the United States; by 1840, this number had increased to 458. In 1900, some 24,600 patents were issued, and in the same year nearly 40,000 people had applied for others.[49] One could cite statistics for the next century, which would simply be larger numbers for more varied ideas. Issued patents generated more facts and publications, if for no other reason than to explain what they were and how best to use them. New products and information improved agricultural productivity, actually spectacularly, repeating the same consequences in manufacturing and process industries.[50] The American workforce doubled their standard of living every forty years since the early 1800s.

The economy grew so wealthy that it could afford additional quantities of information. Economist Robert J. Gordon characterized the prosperous period from 1870 to 1970 as "unique in human history," and in language considered extreme in economic circles, a "revolution."[51] The number of people in the workforce increased sharply, too. For example, in 1840, they consisted of 5.6 million souls, with nearly two-thirds in agriculture. The labor force doubled by the Civil War, while farm shares dropped to 56 percent. Domestic product per capita doubled in the same period.⁵² The size of the economy then took off far more than before, with real GDP growing in the twentieth century despite recessions and depressions. By 1929, the economy had grown in size to $1.1 trillion and ended 2000 at $13.13 trillion. But it kept growing. In 2023, it reached $27.36 trillion, continuing as the largest national economy in the world.⁵³ The workforce had changed too. One number suggests the trend. Between 1950 and 1970, when scholars became increasingly aware of the importance of information in energizing the American economy, the agricultural workforce dropped from just over 20 million to less than 10 million. To put that trend into perspective, the U.S. workforce in 1950 consisted of 62.2 million people, of which just less than a third worked in agriculture. In 1970, both the population sixteen years or older and the total civilian workforce had increased, the latter from 62.2 million to 82.7 million people. Anyone not working in the already technologically advanced agricultural sector was engaged in Second Industrial Revolution economic activities.⁵⁴

The volume of new information available online in computers and then, after roughly 1995, over the Internet caused both the volume and use of that information to expand even faster than occurred with printed materials beforehand. Printed materials continued to increase after the arrival of computing. For example, by 1959, some 15,000 new books appeared every year; that number doubled by 1966.⁵⁵ Rates of increased new titles sped up such that today it has become nearly impossible to track the number of new books, but estimates range from 600,000 to one million per year in the United States. So, there is much evidence to support the finding that many new facts have appeared over the past two centuries.

In *All the Facts,* I demonstrated that there was no corner of American society that did not rely on published and orally transmitted information since the early 1800s. Hundreds of other studies by historians, economists, media experts, and sociologists, among others reaffirmed similar findings. The extant ephemera did too, from old Bibles to Boy Scout manuals, to military training guides, to the vast academic literature so well documented by book scholars. The message was essentially that "everyone" used organized information if they could read. Indeed, one could not perform their jobs or go about their private lives without consulting information that had been put down on paper or was accessible online. Today, making such a statement seems so obvious that it need not be made. That would be a mistake because the historical record demonstrates that prior to the 1840s, while there was much dependence on organized bodies of information, it was less so the case, even less so the farther back in time one goes. Between the 1600s and the 1840s, the world of information had begun to change dramatically.⁵⁶

Historians learned that people relied on information to provide certainty, which the application of scientific methods increasingly offered. Since the seventeenth century, creators of new information honed their skills by creating more rigorous norms for achieving certainty in the quality of their findings, of their facts. In the process, they increased the precision of their findings. All of this was done so that people could thrive and be safe in the world in which they lived. That logic also helps to explain why definitions of information lead to the conclusion that it is a social process, as well as to the findings explained next.

THAT PEOPLE LIVE IN AN INFORMATION ECOSYSTEM RELIANT ON AN INFORMATION INFRASTRUCTURE

While I have touched on the causes of why so much information became available and depended upon, we need to simultaneously keep in mind two other notions: the roles of information ecosystems and infrastructures. They can assist in understanding the volume and use of information on the one hand, while on the other, how that information changed over time. These twin notions further help to explore the possibility that information may be part of a system; in other words, facts play an active role as part of society. Its agency (role) is the subject of much interest, particularly in biology, but it has also attracted the attention of sociologists and other scholars across the social sciences, beginning in the nineteenth century. So, a brief introduction to the ideas of ecosystems and infrastructures is in order.[57] Evidence of how information is being understood and used through the lens of these two ideas unfolds in subsequent chapters. The metaphor of an ecosystem is an old one expressed over the centuries in different ways and language. Information—knowledge—comes from community activities, not just from the certitude of one person, largely because people will not always agree as to the validity of a fact. But when a group—say a discipline or profession—gravitates toward a collective validity of a fact, it is accepted more readily until displaced by another, hence the early realization that a group exists in support of information, a social network. Students of information became interested in ecologies once they realized that humans were not the only forms of life to have intelligence. As the insightful James Bridle observed, people began to realize that "the non-human world seems suddenly alive with intelligence and agency," with people finally taking all of this seriously.[58] He held out the possibility that the entire planet was one massive intelligent being that did things; hence humans are only a small part of a more massive ecosystem.

The earlier of the two notions to attract interest across multiple disciplines is networking. It was conjured up by people's familiarity with telegraphy, and telephone networks, often called "webs of interconnections," later online computing, and most recently, their understanding of what constituted the Internet. Within academic circles, the ideas promulgated by Claude Shannon about telecommunications and Norbert Wiener (1894–1964) and others about feedback loops led scholars and others to understand the concept that nervous systems in mammals (people too) transmit signals (information) back and forth (e.g., my finger has a cut, send help to stop the bleeding, this cake tastes great). These were ideas that spilled out into many disciplines as useful analogies for whatever someone was describing. Since the 1970s, multiple disciplines have embraced the concept of "social network analysis" as a useful intellectual framework for working out theories, shaping descriptions, and developing methods for research and appropriation. A half century later, some have argued that the subject "serves as a central paradigm for understanding social life."[59] Digitization of vast quantities of information after 2000 offered researchers enormous quantities of new data suitable for network analysis.

In its simplest form, networks "consist of points and the lines connecting them," which can be, say, two people (each called a *node*) with a line of communications between them (often called an *edge*). Nodes can be schools communicating with others or with parents and governments. They are also called actors or agents, while edges can be far more than, say, a telephone line, increasingly "relations among nodes" (people).[60] Recall the children's game where two individuals each have an empty cup connected by a string attached to the bottom of each, stretched tight between them. Each child and cup is a node and the string is an edge. Note that, in this instance, the node has three components: a person, a communications device (cup and string), and a role to play: to transmit and receive information. Combined, the nodes and edges constitute a system for moving information (in the form of sound waves) back and forth. Shannon would have argued in the 1930–1950s that these children were transmitting information (*signals*, his word), albeit not as efficiently as telephone technology. He probably played the telephone game as a child.

Much is made of information being organized within disciplines as if they were self-contained, implying that a fortress protects a body of data (facts) from intrusion by people from other disciplines. The networking phenomenon we discuss throughout this book challenges such a mindset. As early as 1935, the distinguished English astronomer, physicist, and mathematician, Arthur Eddington (1882–1944), while discussing the relationship between science and religion, made the point that "the compartments into which human thought is divided are not so water-tight that fundamental progress in one is a matter of indifference to the rest."[61] He made this observation before the availability of such tools for combining information from multiple disciplines as telecommunications, the Internet, and Big Data software. So, the idea of spillover has been with us for some time.

One can think of infrastructures as consisting, too, of physical artifacts and content. With respect to information, the physical consists of such edges as railroads, trucks, newspapers, books, telephone lines, cell phone transmission towers, the Internet, Amazon delivery services, and a nation's postal service. These are required for the transmission of information in whatever manner it exists. Ultimately, information must arrive to an intended receiver in terms they can understand. For a human, this means, say, in English, on a screen or in a book; for a frog obviously in another format unintelligible to humans but not to them. All living matter depends on some physical attributes, even if only to sound waves or chemical transmissions (e.g., among tree roots).

The second aspect—content (information itself)—travels through the physical in the form of information in a book, on a label wrapped around a can of beans, e-mail, charts, spreadsheets, data itself, pictures and texts, among others. Both sustain each other. Without content, there is no need for a physical infrastructure, so no purpose for its development, while content without some way of getting it to whoever or whatever needs it is useless; its existence either impossible to sustain or purposeless. These two components of an infrastructure—physical and content—are central to understanding information. They imply activity—movement back and forth of content—and diversity, since it can be, as social scientists suggest, any manner of interactions. Biologists include systems of nerves, cells, and DNA in their mix of nodes and edges, the physical and content. We are all saying the same thing: information is dynamic and diverse.

The emergence of reliable, more usually accurate information is thus the result of the norms we spoke of earlier in the context of definitions. The behavior, for example, of peers willing to judge the quality (accuracy) of information required an ecosystem (social system) and a process of communicating with each other to move a fact from possibly being accurate or inaccurate to presentation to the public and subsequently to its acceptance. As one student of the process put it: "Over time, as evidence and arguments accumulate, scientists figure out how to resolve impasses, or at least bypass them for the time being."[62]

Every discipline strives to explain all manner of things and ideas within the context in which they exist and behave. Humankind learned eons ago that perspective—context—is crucial for understanding the relevance of information. Historians want to describe why World War I occurred before telling us what battles took place. Doctors want to learn the origins and functions of a virus in the belief that, armed with such insights, they can develop a vaccine or a cure. For several thousand years, theologians in Western societies articulated Jewish, Christian, and Muslim beliefs about the hereafter and what one must do while alive by explaining the context in which both sets of admonitions exist to be acted upon. Since the 1960s, the perspectives of naturalists, cultural anthropologists, biologists, sociologists, and most recently historians (who too speak of ecosystems) have influenced scholars across multiple disciplines. The concept is simple to envision and so serves as a useful theoretical model to explain many activities, both human and otherwise. Michael Polanyi called the community and its network a "Republic of Science," but his idea was perhaps too narrow because the social construct includes everyone, not just scientists or academic disciplines.[63]

Imagine a jungle, say, in Brazil, consisting of various types of trees, fauna, animals, bugs, birds, people, grass, rocks, and dirt, also humidity, water, a particular weather pattern, and so forth. The key idea is that everything in a jungle is there for a purpose to benefit one or more members of its community (ecosystem). Small mammals are there for tigers and alligators to eat, plants and rodents for the mammals to eat, water for all to consume, rain to replenish the water supply. Worms fertilize the soil, which in turn makes it possible for grasses and other plants to grow that sustain the fauna that the deer will eat, before large cats consume them. The point is that in an ecosystem, everything and all living matter are dependent on parts of it to thrive and simultaneously for the whole to do so as well. Ecosystems, like networks, are dynamic because they change constantly (think sunny one day and rainy the next), are incredibly diverse (from small cells to elephants), while the volume of activity evolves, such as with the annual migration of herds driven by changes in weather, seasons, and access to water.

Networks of communication nest in that jungle. Birds warn each other of predators, one deer signals the presence of a tiger, while the roots of a tree share nutrients and water with others. The ecosystem does not function without communication, that is to say, *sans* the flow of relevant and timely information relevant to each resident in that jungle. The more one studies the nature of information, the more relevant the concepts of networks and infrastructures become. In the 1930s, anthropologist Claude Lévi-Strauss (1908–2009) studied indigenous tribes in Brazil as might an ethnographer. This experience made it possible for him to argue that these "savage" people had a profound understanding of their ecosystem—their jungle—armed with much information that was as subtle and sophisticated ("civilized") as what a human today needs to survive the jungle of life in a large city.[64] He is remembered for his work on theories of structuralism and their attendant anthropology, but his perspectives transfer conveniently to notions of information ecosystems. His work contributes to our understanding of the broader role of information within the context of information ecosystems. Historians should use his notions as they work in various circumstances.[65] They help one to understand why there is so much information and its features, regardless of which disciplinary perspective one identifies. In the last two chapters, we return to this observation.

There are social constructs that help define features of such ecosystems to take into account. In the United States and in many European and South American nations, societies formed that allowed for the relatively free flow of information, protected by constitutions, laws, and local cultural practices. Corollary to that fundamental feature of information ecosystems came components such as a society able to afford the cost of creating, storing, and using information, providing economic incentives to create and use more via copyright and patent protections, investing in literacy, and relying on innovations in science, technology, and business activities, which collectively reinforced the centrality of information in such social systems.

As scientists study what happens in a jungle—an ecosystem—they are learning that information is not only housed within the brains of every resident in that jungle, but also collectively within the larger ecosystem. Let two observers explain: "People are like bees and society a beehive: our intelligence resides not in individual brains but in the collective mind."[66] One uses their privately housed information, but that, too, of the community at large, which is stored in other people's bodies, and, of course, in books at the public library and on the Internet. One's collection of knowledge, hence their ability to act intelligently, is the result of distributing information between minds. Some psychologists are explaining that the ecosystem is often more deeply informed than individuals who, by design, keep minimal amounts of information stored in their brains because they can tap the larger world for what they need when searching for more deeply informed knowledge.[67] In short, information is produced and used by a community at large, in an ecosystem. Its purpose is to provide truth, objective reality, because only accurate information is more effective than creeds or beliefs. Information ecosystems facilitate the development and application of "rules for reality."[68] Mutual interactions are the most singular and obvious feature of ecosystems, all fueled by the flow of information.

THAT INFORMATION CHANGES DUE TO EVOLVING NEEDS

The dependence of all participants in an information ecosystem presupposes that as the needs of each evolve, so too will the information one requires at any period in their life, or that they create. A child uses different information compared to an adult with a job. Descriptions of an adult's necessary information demonstrate that dependence on information evolves. For example, the American government has documented the educational requirements for hundreds of jobs since the 1870s, making it possible to track and compare what information and skills, say, a plumber or a teacher needed from one decade to another.⁶⁹ The content of courses taken in high school or colleges and universities evolved—and increased in volume and variety—over time. Uses of infrastructures changed. In 1840, one did not need to know how to use a telephone (as it had not yet been invented), but increasingly after 1900. Three generations later, a working knowledge of how to operate a personal computer was becoming essential, how to access the Internet soon after, and today a smartphone. Yet, old skills were still required, most notably how to read a paper document or a book. In 1840, one did not need to know how to type, because typewriters did not become available until the second half of the century, but today children learn "keyboarding" (today's term for typing). Teaching cursive handwriting is declining, while block writing is in ascendancy.⁷⁰

Changing, adding, and using information is a communal activity involving everyone. Like animals and fauna in a jungle, all people participate in these three activities. Because so many are now aware of this behavior, they think they live in an Information Age. I caution in chapter 3 that they have been living in such an environment for a long time, just that today they are reacting as if they had recently bought a new Nissan Altima. The historical evidence overwhelmingly confirms this behavior. Because information is communal—available to all—and personalized, it varies. For example, a Gallup poll might report that White middle-class Americans who have graduated from college believe the economy is in good shape, while an African American who has not completed high school would reach a more negative conclusion. Both cohorts might have been presented the same empirical evidence (or not), but certainly the same question, and yet to each, the evidence before them could result in diverse conclusions. Since both are certain to act on what they know (or believe), the White American might seek employment and opportunities for career development, while the other might conclude that such initiatives are less available or simply not the same.

Scholars are learning more about ever-expanding groups of people using information. Already mentioned is the study of what academics have done over the past two centuries, the results I reported in *Birth of Modern Facts*, and the growing interest in what appear to be similar behaviors among artisans and craftsmen (i.e., blue-collar workers and others in the "trades"). The language used to describe the changing nature of information also evolved. Scholars used the word "knowledge" a great deal until the mid-twentieth century, earlier "skills" often to mean the same thing, then afterward "information," "data," "signals," "noise," "bits," and "bytes." Considerable attention paid to the creation of new bodies of information expanded the scope of study, such as how these grew in disciplines and occupations, each with its own cadence of change and evolution.⁷¹

The usefulness of information goes far to explain the need for changing facts. As the need arises for new information to, say, solve a problem at work or to understand why a particular phenomenon in physics plays out as it does, people hunt for new data, information, and explanations. As those prove effective, appropriation increases, because people want to be productive economically, comfortable in understanding how their world operates, and to be safe and prosperous. These are normal aspirations that help explain the evolution of societies and species. The survivability of residents in a jungle as a whole—society/ecosystem—itself depends upon such behavior by all. Academic disciplines study the role of information in this way. Sometimes the language used is economic, in other circumstances it concerns technology, and in recent decades a combination of both. Economists look to both as sources of increased labor and general economic productivity. Two historians explained the use of the term "knowledge," "as a proxy for the entire gamut of inventive and innovative attitudes about work organization and information flows in industry, the state, and other communal forms."[72] They are not alone in combining the two topics.[73] The emphasis comes back to our earlier proposition that information was created and applied when it was practical to do so.[74] That is why, for instance, new physical components of a network were created—because they worked. It is why new bodies of medical information were created because they did, too, to cure diseases and heal the injured. This behavior proved universal, not limited to the United States, parts of Asia, or Western Europe.

The notion of an adaptive cycle increases our understanding of the practically evolving nature of information. It emerged as a byproduct of studying what happens inside ecosystems. As an intellectual tool, it is used to understand how the processes of destruction and reorganization take place to keep linked together analyses of how systems are organized and maintain resilience while undergoing change. The adaptability of an ecosystem—hence the information attached to all its parts and participants—can help students of one, or of information's transformation, understand periods of change and others of stasis where activities and facts remain stable or less subject to transitions to new forms. This has allowed scholars to see ecosystems as going through phases of growth, exploitation, conservation, collapse or release from prior conditions, and reorganization.[75]

This set of four functions represents a partial departure from the cycle of change and rupture that Thomas Kuhn proposed occurred over a half century ago. Pressure builds over a period of time that results in some change or reorganization that occurs in a shorter time. We may have seen that happening with information, in that historians have argued more change came in its types and volume since the Renaissance than in the 1,500 prior years, more so since 1750, and yet even faster since the 1870s. Adaptive systems do not function in isolation; they nest in prior activities and bodies of information that influence the surge in new knowledge that occurs from time to time. For example, information that made possible computing emerged from prior knowledge of how electricity travels and behaves. That appreciation was reinforced with a growing understanding of its physical components (e.g., types of wires, switches).

Why do we care? Because the widely held notion that information simply piles on top of earlier bodies of facts, suggesting society progresses continuously in its accumulation of knowledge at some pace, is in reality challenged. Book historians may talk about the increasing number of volumes produced over time, universities brag about how many books they have in their libraries today versus in some earlier period, and technophiles about the amount of information available on the Internet. But none of these types of facts are necessarily insightful, just easy to collect. Furthermore, online databases close, while librarians shrink their collections of books and discard paper journals all the time. Ecosystems and their information change collectively at different speeds and at the individual participant's level. Case studies of disciplines and individuals within them reflect much of the behavior called to our attention by adaptive cycles.[76]

One challenge facing historians of information's evolution is how to deal with biases (or assumptions) that they bring to the issue. Historians are not unified in their suppositions, accepting a diversity of opinions, which they view as healthy for the winnowing and sifting of findings and facts. They, like all practitioners in other disciplines, have embraced the values and uses of scientific methods of research and presentation.[77] Historians think of events as unfolding in circular repetition, linear improvement, and, to use Rosalind Williams' phrase, "rolling apocalypse."[78] Of course, all three are logically incompatible but, to quote the same historian, "existentially they are complementary."[79] Historians summon one or another when it suits their purposes as useful mental models.

The first speaks to the repetition of human activities, the second to the notion that progress ensues as we learn more and thus become increasingly safe, prosperous, and possibly happy. The latter worldview has dominated thinking about the evolution of science and technology. However, the third raises the specter of uncertainty and unpredictability, not just obvious problems, such as global warming, the possibility of nuclear war, or the "Mother of all plagues." It is the issue raised by scientists and mathematicians about now always knowing realistically about a fact. Students in the 1960s were told that it was a fact that there were nine planets; in the twenty-first century, Pluto was kicked out of the planet club and now nobody knows how many there are in space. Progress in what is known and uncertainties coexist.

One additional feature of changing information crucial to understanding its evolving nature is the influence of precision. It is usually discussed within the context of mathematics and numbers, the former as a language, the latter as a more exact statement of a fact. Mathematics and statistics came most widely into their own across multiple disciplines and practices in the past 175 years as tools with which to work with numbers.[80] But precision meant more than numbers; it involved increasingly specialized information. Having a body of information about how to grow corn or wheat using "scientific" methods in the 1870s was fine, but by World War I, farmers wanted, and indeed needed to know, how to grow those same crops in Nebraska versus in Virginia, in central France as compared to Spain's Andalusia. Across the past two centuries, the move to numeracy was nothing less than staggering, fundamentally altering disciplines and professions and their bodies of information, notably in the hard sciences and economics, increasingly in the social sciences, and in public administration. It became standard form that most insights in the hard sciences and engineering had to be expressed mathematically, and for good reason, because numeracy added insights otherwise not attainable by observations or the use of adjectives. Accounting, machining, computer chip manufacturing, medical diagnosis, aviation, and such trades as electrician, carpentry, road construction, and heating and cooling experts all embraced precision as both core values and essential activities.

Every recent generation of students has been indoctrinated into the power and use of statistics and mathematics. In 1908, two textbook authors advised teachers to make sure every student learned these skills; otherwise, that poor soul "must ever remain a plodder, a waster of time, and a blunderer," living a life without access to the newly emerging bodies of information required to thrive in the American ecosystem.[81] This mantra permeated modern society, giving these subjects a patina bordering on divine revelation in the twenty-first century. However, we have occasion in subsequent chapters to see some pushback as the need to synthesize so much new information in so many new disciplines exists, such that there is a growing call for the return of the adjective, of a well-written narrative description of a body of facts. Wisdom may see a surge in popularity, this time buttressed by the use of digital tools and not simply through the accumulation of experiences by aging individuals. But let us not mistake the broader influence that all manner of education has on the use of information, whether precise or not. As Alexander Luria, considered the father of neuropsychology, pondered, while probably not thinking specifically about mathematics, statistics, or precision, his observation from the 1970s applies ever more so today. He is worth quoting at length: "Education, which radically alters the nature of cognitive activity, greatly facilitates the transition from practical to theoretical operations. Once people acquire an education, they make increasingly greater use of categorization to express ideas that objectively reflect reality."[82]

The combined need for precision and information relevant to a particular problem or issue leads to the logical question of why so much of it has come more from some disciplines than others, especially since the mid-nineteenth century. People with similar interests and skills obviously flock together, representing one reason: clusters of like-minded people work more efficiently together in collaboration, or gather to learn from each other. But a second explanation ties into the notion of ecosystems: disciplines represent communities, think neighborhoods within our metaphorical jungle. They establish their own ways of thriving in that ecosystem. Over time, they work out how to produce information relevant to them.

The economist Benjamin M. Friedman explained:

> As an intellectual discipline matures, its conceptual core normally becomes less subject to external influence—from worldly events, from thinking in other areas of inquiry, from the culture of the day. Fundamental thinking within a mature discipline more and more tends to follow its own momentum, while the role of such outside forces becomes increasingly a matter of application and method.[83]

However, he argued persuasively that religious values also influenced, at least, the fields of economics and business practices. We have occasion in this book to argue that (a) disciplines had their ways of creating and using information—a finding of the prior studies buttressing this book's observations—and (b) that recent developments are either fracturing disciplines into ever-narrower ones (e.g., in biology) or are becoming more able and willing to adopt practices, data, and values from others. Both patterns of behavior are in evidence operating simultaneously in the early decades of the twenty-first century.

One can detect the whiff of stability as a desirable feature. We may learn someday, however, that the laws of thermodynamics affect the nature and role of information too, exuding strong odors of entropy and further specialization. I suspect that we will need to experience extensive use of AI for several decades before we can identify what is happening. The analogy of the jungle would suggest that both whiffs and odors will remain in dynamic forms.

THAT BIOLOGISTS ARE SHAPING OUR UNDERSTANDING OF INFORMATION

Biology has a long history but became an identifiable field of study in the nineteenth century and a broad collection of subfields in the twentieth. All living matter became the purview of biology, hence connected to the study of ecosystems. In the nineteenth century, biologists, medical researchers, and others identified the existence of germs, while the study of chemistry made possible medicines that could cure diseases and treat other illnesses. Since most of what biologists and chemists studied could not be seen by the naked eye, they developed a raft of tools to assist, most notably in the late 1500s and early 1600s, the microscope followed by other facilitative techniques and instruments. The emergence of theories of evolution that helped biologists put into context the transformation of living matter became a second development in the 1800s. A third was the early and extended use of statistics, later formulas and mathematics, with which to study and document their findings. Like other STEM researchers in both centuries, biologists created massive bodies of information—they frequently use the word data—and did not hesitate to use computers to assist in the project, beginning in the 1960s. A fourth development came in the mid-twentieth century: the study of genetics which today is profoundly influencing how biologists view life and humans.

In sum, as historian of brain science, Andreas Killen, pointed out, biology went from being about one's fate to being "the target of medical or chemical intervention," with the brain now being an evolving organ holding out the potential for major medical and other intellectual possibilities.[84] One byproduct discussed later in this book is the resurgence of interest in all matters concerning human, and later non-human, intelligences, including AI. This is an important development because in the late nineteenth and early twentieth centuries much controversy swirled around how smart various racial and ethnic humans were, with the result that in the second half of the 1900s discussions of intelligence (e.g., IQ tests) were fraught with social, cultural, ethical, anthropological, and political baggage; far less so in the subsequent century. Between growing recognition that intelligence (hence use of information) existed in non-human living creatures and plants and computer-originated intelligence (AI), the topic was back front and center, hence discussed in subsequent chapters. Had this book been written a half century earlier, your author might have worked mightily to avoid discussing such matters. The biochemical professions plowed ahead, picking up where they had left off before World War II.

The results on the information front were prodigious. Each subfield resulted in the creation of massive collections of information. In biomedical fields, researchers publish in excess of one million papers *each year.*[85] Data sets increased from hundreds to billions.[86] Medical literature databases expanded, too, such that by 2020 in excess of 26 million citations existed, covering all aspects of the life sciences and biomedicine, nearly 30 million just two years later.[87] A similar story could be told about the informational activities in chemistry in the 1800–1900s, now superseded by the work of biologists. It is no wonder that biology has much to work with and that its practitioners are a busy lot.

Current biology, specifically genetics and bio- and medical informatics, also combines practices, tools, techniques, and bodies of information, making it possible to study the building blocks of life itself, even examining chromosomes in cells. Historians have documented a great deal of the early history of genomics, so we can skip more to the present.[88] It is also a growing part of the history of Big Data and massive computing. Over the arc of nearly two centuries, chemists learned how to purify compounds, while biologists came to understand the effects of these on people and animals; medical doctors did too. As in so many other disciplines, all encountered the influence of electricity in their understanding of information. As early as the 1700s, researchers had concluded that electricity energized nervous systems, not just chemical substances. Electrical activity became an important subject of interest in observing the behavior of cells, genomes, and DNA. The use of computers and the conceptual understanding of how information worked within information technologies began to inject new notions into genetics and biology as a whole.

That introduction resulted in a new subfield called bioinformatics (BI) in the 1980s. It also reflected a new trend in the use of information evident only toward the end of the century: the merger of data from multiple disciplines, such as from medicine, biology, computer science, information technologies, mathematics, and statistics. This trend toward converging information represented a turn away from the practice of the prior two centuries of researchers developing their own languages and collections of data. Yet the BI community, too, practiced the long-existing habit of establishing their own subfield. In reality, they were also doing the opposite, made possible by electronically based datasets and software facilitated by computing. Using political historical analogies, they were Balkanizing their pools of information while also participating in a disciplinary European Union. Now one began to see the possibility of learning from other disciplines with new ways to think about and manage the study of large bodies of information. AI holds out the promise of facilitating that convergence in the use of information. As one group of researchers exclaimed what became possible: "development of an ever-increasing number of bioinformatics tools" and shift "in complexity, evolving from a gene-centric perspective to the systems level," with computers reducing large collections of data then converting these data sets "into usable knowledge."[89]

Increasingly, biologists began thinking of the nervous system as an informational topic and how this system interacted with the brain.[90] Think telecommunications, real-time transmission of information back and forth through this network, with all cells in, say, a mammal (e.g., person) connected online with the brain. That worldview was reinforced by the use of computing, as biology became "a data-bound science, a 'science' in which all the data of a domain—such as a genome—are available before the laws of the domain are understood."[91] As in other STEM disciplines,[92] biologists began to think of their findings as statements of probability and constraints in how life functions, relying less on laws—"rules of the road"—since few activities in the mind, nervous system, or life are never exactly the same, just as waves pounding the beach are never exactly either.

Just as many disciplines began using the language and analogies of computing, so too did they do the same in appropriating biology's language and paradigms. Two examples will have to suffice to illustrate the point. Jonathan Haidt is not shy about borrowing from multiple disciplines. He is a social psychologist who teaches ethical leadership in a business school, so he studies the psychology of morality. Two centuries earlier he might have been labeled a religious figure, later a business professor, but today a psychologist. I would include in his intellectual baggage systems thinking and networking. But he has relied on biological insights too:

> We should see each individual as being limited, like a neuron. . . . A neuron by itself isn't very smart. But if you put neurons together in the right way you get a brain; you get an emergent system that is much smarter and more flexible than a single neuron. . . . If you put individuals together in the right way . . . and all individuals feel some common bond or shared fate that allows them to interact civilly, you can create a group that ends up producing good reasoning as an emergent property of the social system.[93]

Two others from the same profession draw upon biological understanding and examples to enrich their discussion of information:

> People are like bees and society a beehive: our intelligence resides not in individual brains but in the collective mind. To function, individuals rely not only on knowledge stored within our skulls but also on knowledge stored elsewhere: in our bodies, in the environment, and especially in other people. When you put it all together, human thought is incredibly impressive. But it is a product of a community, not of any individual alone.[94]

Both quotes also parallel thinking about information ecosystems. In short, biological metaphors and analogies are mixed with similar thinking in other disciplines expressed in the parochial language respective to each.

Are we also seeing the slow decline in the power of theories as chance and probabilities influence how one views scientific information? One highly respected scholar suggested a new perspective: "the universe is fundamentally composed of data."[95] In that paradigmatic construct, physical matters are manifestations of a more primary consideration—information—so facts become objects (Shannon's implication), reality's most fundamental component. Biologists now speak in the language of information, such as when they explain how a medication interferes with messages sent to cancerous cells to explain how it works to cure, arguing that the behavior of a cell is thereby altered. Perspectives brought by biologists to their disciplinary shades are spilling over into the social sciences, media studies, library science, and back too, to the hard sciences where much of their original thinking about how information works originally surfaced. All see information as in constant motion, dynamic, and sufficiently measurable as patterns of behavior of practical value to humans. As one observer noted, "We look hopefully to nature to teach us how to do what living systems accomplish with such skill—learn, adapt, and change."[96]

Before leaving the biologists and their related concerns about how humans process information—about which I have much to say in subsequent chapters—one can ask briefly how humans consume information. Scholars who study what they normally call "science communications"—a subfield complete with its own courses, degree programs, journals, and other publications—discuss and train future scientists and the public at large on how to communicate complex scientific ideas and facts. Most people want to make science- and technology-based decisions, while scientists want the public to trust science, and so the challenge is how to make otherwise difficult ideas both engaging and relevant. One of today's answers is to resort to a timeless communications technique: storytelling. Conjure up conversations around campfires thousands of years ago, and even today, and we realize quickly that telling stories may be one of just several very obvious features of humankind that distinguish people from all other living creatures. Stories make possible the integration of multiple streams of facts, situating these into contexts relevant to listeners, and permit wrapping these narratives in emotions using language accessible to those who are not scientists. All three realities affect how people take in information, providing meaning and relevance to an audience that they require before accepting and then using information.[97] Storytelling becomes part of the process humans use to integrate and make sense of multiple bodies of information; they did not need AI to link facts together.[98]

However, students of communications identified three gaps created by the early twentieth century between the role of opinions (possibly in decline when not enveloped in hard data) and the increase in scientific/mathematically based information. These are somewhat obvious: gaps in information between how much experts in one discipline knew when compared to that understood in others; an epistemological gap distinguishing scientific reasoning from, say, opinions; and less visible, moral gaps within learned communities and the public in which the former aspire to tell the truth (i.e., facts of the matter) and a larger public more interested in accumulating power and profits, or leveraging political initiatives.[99] Science communicators desire to fill these gaps, hence all the discussions about the power of storytelling, communications, and the use of AI, for example. Filling the gaps is part of the process by which humans want to somehow gain control over an entire universe of information so that they can apply that which is more useful to them as a society, discipline, or individual. Keep these various issues in mind as we move through a broad set of conversations about the nature and role of information in our time.

Historian Peter Burke should have the final word here because, while much of what is discussed in this book is about the consequences of increased specialization, an issue he acknowledges as relevant, it is not the only one that makes information less accessible: "It is also a consequence of the increasing remoteness of scientific experiments from everyday life."[100]

THAT TECHNOLOGIES SHAPE INFORMATION, TOO

The problem is that the definition of technology keeps changing.[101] In the nineteenth century, one would have thought of technology as something physical or based on mechanics or physics, such as steam, steam engines, and trains, later applied physics and engineering in the form of telegraphy, telephony, and so forth. Machines embodied notions of technology, interspersed with the adoption of principles of science applied to some object or task, such as inventing airplanes, rockets, and atom bombs. Eric Schatzberg, a leading historian of the concept of technology, is not one to mince words: "the definition of *technology* is a mess" and so "the term sows confusion."[102] Most people think of it as computing, opinion writers about anything to which they put their minds. Academics in different fields hardly do better and possibly worse because they use the term to describe how things are done within their respective disciplines. Schatzberg is correct in arguing that "definitions are so broad as to be almost useless, covering everything from steel making to singing," or the application of science, when historians believe scientific input is just that, one of many possible components of technology.[103] Because scholars are increasingly taking a stance that technologies are practices too, they are rapidly promoting this version of the definition as an extension of earlier ones. Businesses are simultaneously trying to patent processes as technologies, causing officials and lawyers to scratch their heads as they decide how to proceed with such applications. We need not engage in the debate as Schatzberg has done a fine job in pulling together the cross-disciplinary discourses over definitions.

For our purposes, I recommend that we view technologies within the scope of information discussions as tools useful for the collection, organization, storage, assessment, and use of information. That approach conforms nicely to what the general public would have thought of the idea over the past two centuries. The word *information* can mean whatever the reader wishes, with my being confident that they understand what it means to them and how to get to it. It is the "getting to it" in addition to storing and organizing that is relevant here. Therefore, technology and its influence on information imply something sophisticated. A database in a computer is more sophisticated than, say, a bookshelf, a keyboard attached to a computer more than pencil and paper. Of course, everything is relative, so a shelf of books is a far more advanced (i.e., technical) solution for the organization and accession to information than, say, clay tablets or rolls of parchments.

The single greatest technological innovation of the past two centuries affecting how information is collected, organized, and used is the application of electricity to all manner of tasks. And the computer? Without harnessing the use of electricity, we would still be using slide rules and mechanical calculating machines. Our notion may seem too simplistic, but better to have a view of an entire field than be too focused on individual blades of grass. Thriving in our ecosystem requires the "big picture" before dealing with specifics. By following developments in the form and use of information, historians have already contended with telegraphy, telephony, radio, television, satellites, smartphones, personal computers, mainframe computers, digital coffee makers and doorbells, software, AI, expert systems, databases, programming languages, spreadsheets, hand calculators, even earlier with electrical desktop calculators and adding machines, for a century with traffic lights, now sensors of all kinds, robotics, automation, and increasingly with electric cars which are large real-time Big Data analytical machines and batteries on wheels needed to keep one's car from crashing while getting us to our destination. But the reader knows this, too. Just keep the obvious in mind—electricity changed a great many things, including information. Many "things" also acquire computational capabilities that require the use of information, such as one's bedside clock or their programmable coffee maker.

For one thing, electrified data processing—computers—could collect much data with or without human assistance, so that findings, research, and instructions are now based on far more facts than were accessible even a century ago. For another, electricity made it possible to quickly add, mix, and match ever-larger bodies of information cost effectively massively beyond capabilities in the pre-electronic era. It was electrification, among other innovations of the past two centuries, that led people to broaden and confuse definitions of technology. No matter, because more importantly, most information was converted into electrical formats and was moved about using the same source of energy. Today, more information is in digital form than on paper; more is on the Internet collected and used by electronic devices than by human beings. Now one can begin realizing why AI has a potentially profound role in front of it as the most ravenous consumer of information ever encountered by humans. Technologies are now using information; no longer do living beings monopolize the information ecosystem. Intrusive species of new technologies are coming in and also using the same energy source to engage with information. Biologists could weigh in to argue that other living forms consume larger amounts of information, but I promised to keep this book just anthropomorphic.

The availability of computing, in particular in formats humans could use that were affordable during the past century, can be seen as a corollary to the statement that electricity made possible a more profoundly changed nature and use of information. In quantum fashion, computing facilitated the collection and use of information far more diverse than any other prior method could. Computerized manipulation of information begat more manipulation. It is an old, well-known behavior, often called the "turnpike effect." If you live on a small road that takes 45 minutes to travel to work at 25 miles per hour, a commuter makes one round trip a day. If, then, the town converts it into a four-lane highway with driving speed permitted to be 55 miles per hour, one needs only 20 or so minutes to make the trip. The commuter might now think about driving home for lunch or to let the dog out. Singlehandedly, that person will have doubled their own travel because of the throughput convenience offered by the new highway—two round trips in one day. The same works with communications and computing. Today, when writing a book, I can quickly go to the Internet dozens of times in one day to confirm someone's name, birth and death dates, and complete an endnote citation, while thirty years ago, I would have accumulated research points and trooped down to my local university library to get all this work done in one trip. So, we all use more information. We have gone so far that even mainstream publications run voluntary tutorials on how to do data analytics for we mortals.[104]

The turnpiking effect, of course, has consequences. A highway can divide a community if it runs right through it, as do some in large American cities, while the number of traffic accidents increases. On the other hand, one can now take a better job farther away from home, operate a package delivery service not practical before, and engage in other activities. Electrified information subject to the turnpiking effect encouraged people to make more of it in greater varieties applied in ever-more novel ways. Computing facilitated a process already underway of specialization that began before electrification, now speeding it up. It would be difficult to imagine the amount and variety of information we have without computers. So we can conclude that computerized handling of information changed it and caused those to be directly complicit in all these transformations that occurred to humans and their societies in the past century. We are at a point where people complain about too much information, or ineffectual data.[105] The process began long before the arrival of the Internet, of course, and affected how books and magazines were put together and offered to the public, while older methods for handling information remained. We still read paper books, write diaries (blogs too), and insist that our university libraries preserve "civilization" by filling buildings with papers and books.

What biologists are beginning to contribute to our understanding of information is that facts—information, data—are active agents operating in both physical environments (e.g., in cells and in all living matter) and in the social constructs that we call societies, nations, neighborhoods, and families, and that for convenience one can think of as held together by information ecosystems. The existence of information infrastructures within those ecosystems provides the actual information needed to sustain an ecosystem, as well as the wherewithal for moving it about. The notion of networks provides a convenient metaphor for understanding two ideas demonstrated as existing, thanks to those who have explained how electricity works and how messages are moved about, say, along an electrical wire or inside a computer chip. It seems data is not static, as it is in constant motion and evolves as it moves about. Living matter uses and controls it when possible. That is why an information ecosystem is constantly active. To observers of how information is created, organized, and used, it is understandable why some might go so far as to say that information is life itself, that it is our reality. We may need more convincing about that centrality of information, but we live in a period in which biologists are having their turn in making the case.

Taking advantage of the "mess" Schatzberg called to our attention, one can account for yet other technological/physical influences on the nature of information. Those who study material cultures, particularly operating in scientific subfields such as material sciences, are increasingly engaging in debates with physicists, chemists, and biologists, serving up perspectives that tie neatly into notions of information ecosystems. Mark Miodownik has argued that,

> living matter is, in some sense no different conceptually from non-living matter. What dramatically distinguishes the two is that in living materials we find there is an extra degree of connectivity between different scales: living materials actively organize their internal architecture. They do this by setting up communications between the different scales of the organism. In a non-living material, a mechanical stress imposed at the human scale has all sorts of effects at different scales, causing many internal mechanisms to react in response: as a result it might change shape or break or resonate or stiffen.[106]

Living matter experiencing the same exogenous forces on it detects this activity and adopts a course of action in response. His intention was not to denigrate the activities of living matter, but to "upgrade inanimate materials," as traditional disciplines discover in combination that "they are more complex than they appear."[107] Regardless of the role played by computing technologies, the arrangement of living and non-living materials on earth remains unaltered. It is a bedrock belief of scientists in all disciplines. Things are alive or not, animals vs. rocks, computers, and buildings. Even in this dimension of material sciences, disciplinary boundaries may be converging, as we glimpse today with early developments of synthetic organs and possibly AI.

But what about today's big elephant in the room: the Internet? It is a worldwide envelopment of Earth, physically through satellites and electronic signals traveling over the physical space above the world. Increasingly, it is being viewed across many disciplines as a combined system of physical, social, and information communications. Its activities are reflecting the kind of dynamic actions biologists have spoken about for decades: information moving about, interacting with other information (the idea of feedback), structured like nervous systems in living matter, and as a social construct influencing new behaviors in human society, such as stimulating the diffusion of fake facts, conspiracy theories, and interesting information presented in a timely fashion to humans (e.g., Google-delivered facts to settle a bar bet) on a massive scale by human standards. Digital and analog sensors communicate with each other over the Internet without interventions by people and with AI in the form of learning systems that draw lessons from such communications to make decisions about future actions.

The Semantic Web—as this activity has long been associated with notions about the Internet—is now a reality. Students of information and how it fits into human activity increasingly need to understand the concept behind it. Let an expert explain:

> Today the Web embraces both data and automated functionality as part of its make-up. Web Services, a recently introduced Web-friendly set of technologies, allow back-end computer systems, programs, and processes to interact using the Web as their door to door courier of data. Furthermore, multiple layers of data now exist on the Web, and so it is not uncommon to find cases of data about data, data about that data, and so on.[108]

Lest anyone question the reality of his statement, he wrote that description in 2007, so if anything, his explanation has become more an embedded truth than might have been the case nearly two decades ago. Philip Tetlow, the expert just quoted, reflected on the Semantic Web as a system that "points to a self-reflective world of information that freely talks about itself across multiple levels and dimensions. These dimensions not only add volume and weight to the descriptiveness of the Web's data, but also open doors to far greater levels of computerized automation."[109]

This capability is what made cutting humans out of many communications possible and now an important, indeed increasing, possibility. In other words, the Internet is slowly beginning to act like a living organism.[110] That the Web is hardly visible as tangible things, it exists, nonetheless. A page on the Web can be linked to another regardless of whether it is in the same file on your computer or halfway around the world. It works at faster speeds, too, than humans, and interrupts its own cadence of communications when it deals with humans who, so far, are allowed to come in and off the Web as their choice. Since life is usually described in terms of its ability to communicate, that is to say, transmit information through various systems, such as with its DNA, nervous system, and cognitive activities (including talking), is it possible that some technologies are becoming more lifelike?

SOME CONCLUDING THOUGHTS

Voltaire raised a point still unresolved but worth keeping in mind. By saying that asking questions is more than simply settling for some definition of information, we are compelled to disturb preconceived equilibriums. My notion of information may not be the same as yours, so this difference begs the questions: Why? How so? With the passage of some 250 years since Voltaire pondered such issues, it appears he was right. We have learned so much about information—what it is, how it moves about, its role in society and economies of the world, and how it can be appropriated by computing and AI—yet it still is subject to laws of physics and, more tactically, to how electricity behaves. We run into the potential danger of believing in some informational determinism, as did many earlier scholars who argued that computer technologies had done just that.[111] They are now being admonished to avoid such a narrow perspective of some digitally dominant view of society.[112] Biologists and those who study the functions of brains, nervous systems, and DNA could form a cabal with the earlier electrical engineers who explained how telecommunications worked. It is a risk—and potential opportunity—we take in studying information. Recent AI and robotic developments have once again raised the fear of digital fascism—the technological determinism that some feared would dictate how humans went about their lives.[113]

What one can take away, so far, is that information is not just abstract, a construct of the mind. It is real, moves about, and influences the behavior of all living matter. Sticking to the commitment to remain narrowly focused on human relations with information, the next several chapters add perspectives about how people used it to work, how their views about the role of information in their lives affected their application of facts to live in their ecosystems, followed by their ever-increasing use of data made possible by information technologies. A parallel relevant discussion about the quality—truthfulness—of information as it exists within an information ecosystem is essential today given the recent uptick worldwide in the influence of fake facts. These chapters remain more rooted in the immediate past. We begin with what drove so much change in modern times: how information and computers changed work, building on research conducted over the past three decades by many historians, including myself.

Finally, we should acknowledge that modern society, say, from at least the eighteenth century to the present, benefited economically to a sufficient extent to afford the creation and use of new information. There is an economic underpinning that cannot be ignored. Economist J. Bradford DeLong, like most economists, celebrates the plethora of inventions that made modern society possible, but he pointed out something just as important: "the systematic invention of how to invent," including not just ever-larger enterprises and government agencies involved but also "organizing how to organize." He argued convincingly that both features of modern information development and use were required.[114] My only caveat is that the modern processes of organizing to create information have been underway since the 1600s, if not earlier. We are just now, however, coming to realize that fact.

How all that happened takes us to the subject of the next chapter. There, we explore how information and its modern technologies changed the nature of work and created new social structures.

Chapter 2

How Information and Computers Changed Work

> *The only irreplaceable capital an organization possesses is the knowledge and ability of its people. The productivity of that capital depends on how effectively people share their competence with those who can use it.*
>
> —Andrew Carnegie

This chapter argues that all industries, the majority of companies, and government agencies used computers to such an extent that this technology and the information it handled profoundly changed how work was done in the second half of the twentieth century. Like librarians and members of other disciplines, business professionals struggled with information overloads that created the need for information technology (IT)-centered solutions for the organization, storage, and use of information as these became available. The influences of constituencies reported in my earlier book, *Birth of Modern Facts*, and by other students of the use of information are reflected in society's massive appropriation of computing.[1] I argue here that how managers and their organizations operated changed significantly over the past seven decades. I describe what that transformation looks like. Their roles help to explain why computing, indirectly information theory and technological realities, connected so many activities and applications beyond the "telephone network" thinking of the 1920s–1960s. This chapter explores how such efforts manifested themselves in the broader world of jobs and economic activities. I discuss the uses and consequences of using information, taking the story into the twenty-first century.

It is useful to understand the extent of the transformation that occurred at a high level. Using the example of the United States, innovations resulted in both the requirement for more extensive uses of business information and the need for workers to have more skills requiring their use of facts. Table 2.1 displays data on how occupational groups changed their percentage of their shares of the total workforce between 1860 and 2015. It shows at a high level the shares in 1860 versus 2015 by occupational groups, representing nearly 300 different types of jobs. Leaving aside the fact that the total size of the workforce increased over the period, and just concentrating on the share changes, demonstrates that "knowledge work" clearly increased as a percent of the total. For example, office administration, which accounted for half of one percent of the total workforce in 1860 expanded to 12.4 percent by 2015. That rise in job share reflected an increased demand for people who handled information, in the beginning on paper, later additionally with telephones, telegraph, radio, tabulating equipment, and finally computers. In the same 155 years, production workers shrank from 14.2 percent of the workforce to 6.5 percent, even though, as with office administration, the actual total number of employees went up.[2] These trends for production workers tell us, too, that automation and computing profoundly increased productivity such that manufacturers needed fewer of them to do even more work in the mid-to-late 1900s than for similar volumes of work a century earlier. To provide brief context, table 2.2 lists how many employees are compared and the size of the economy in both years.

Other job categories reinforce the fact that as work became more information intensive in an economy that as a whole relied on more information, changes in percentages were high. Management increased its share of the workforce to 8.4 percent, only surpassed by their employees, the office staff, who increased their share from 1860 to 2015 by nearly 12 percent. Healthcare, another information-intensive profession, grew its share of the workforce to 8.5 percent of the total, but more interestingly, moved from 1 percent of all workers in 1860 to nearly 10 percent in 2015. STEM communities increased their collective share to over 7 percent. Growth rates occurred

Table 2.1 Changes in Occupational Structures in the United States, 1860–2015

Occupation Group	Share (%) 1860	Share (%) 2015	Change in Share (%)
Administration	1.0	16.3	15.4
Management	4.8	13.2	8.4
Healthcare	1.0	9.5	8.5
Sales	2.5	8.7	6.2
STEM	0.1	7.4	7.3
Production	14.2	6.5	−7.6
Laborers	9.6	3.6	−6.1
Agriculture, Fishing	43.2	1.2	−42.1

Note: Does not include all groups, such as construction, mining, transportation, teachers; objective is to provide a sense of the shift over time. Source: Modified from Joel A. Elvery, "Changes in the Occupational Structure of the United States: 1860 to 2015," Economic Commentary Number 2019-09 (Cleveland, Ohio: Federal Reserve Bank of Cleveland, June 26, 2019): 3.

Table 2.2 Number of Workers and Gross Domestic Product (GDP) of the United States, 1860 and 2015

Year	Number of Workers (millions)	GDP (current dollars)
1860	11.1	$5.4 Billion
2015	158.0	$18.22 Trillion

Source: U.S. Department of Labor, U.S. Bureau of Economic Analysis.

among teachers, professors, and librarians, while jobs that required relatively less information shrank as a percentage of the total, such as the category called Laborers. But even those employees walked around with smartphones and laptops in 2015, so their dependence on information to do their work was *relatively* less, not less than a cohort from 1860.[3]

The economist who assembled these statistics, Joel A. Elvery, pointed out the undercurrent of the transformation, setting up the main observation made in this chapter: "When a business adopts new technology, produces new products, or makes other changes to its processes, doing so usually causes changes in occupational mix." He adds, "Consider what happens when a factory automates. The number of production workers usually declines, while the number of engineers rises and the front office stays stable."[4] While the share of production workers declined, those in management, engineering, and office work went up. It happened in service industries too. For example, when banks installed computerized check processing in the late 1950s and early 1960s, the number of clerks who used to do that work declined by some 20 percent, which was about the same percentage as happened to staff doing inventory control in manufacturing.[5] Elvery observed that the greatest churn occurred prior to 1970 and that subsequently the relative share of each occupation remained stable.[6] Changes after the 1970s occurred at half the rate that they did in the most turbulent period of the 1900s through the 1940s.[7] He was concerned about how dynamic the economy was (or not), while the data shows a consequential transformation in how work was done due to new information and uses of information handling technologies.

This chapter focuses on the theme of the diffusion of information through organizations by way of computers. Every form of information spread through enterprises and economies with the assistance of computers, increasing the amount of facts that any individual or institution had access to during the second half of the twentieth century. Because of the central role of computers, I have organized each section below largely by the major types of computing that became available in approximately chronological order. Each technological innovation made possible new forms of diffusion, and entire industries did not hesitate to exploit these. To set expectations, this chapter is not intended to be a history of computers; our focus remains on information.

EARLY USES OF COMPUTERS, 1950–MID-1960s

In the half decade following World War II, companies and government agencies in the United States and Great Britain became aware of the emergence of this new technology called computers, Western European organizations in the early 1950s.[8] Companies were the first to use computers on a large scale for routine information handling activities, closely followed by government agencies, both beginning in the early 1950s.[9] Commercial uses—called *applications*—became possible when mainframes became reliable and affordable. Computing began to mature in the late 1940s when government agencies funded experimental machines, largely for use in military and scientific applications.[10] But the path to commercial use came with the ENIAC, manufactured by the Eckert-Mauchly Computer Company (years later part of Univac). Soon after, American electronics and office appliance firms entered the new market, notably GE, IBM, NCR, Burroughs, and later RCA.[11] In the early 1950s, computers were seen as systems, just as tabulating machines had been in earlier decades. These new systems consisted of the computer (central processing unit), and for input, punch cards of old, magnetic tape by the mid-1950s, and beginning in the second half of the decade, direct access disk drives. Output consisted of the same devices plus printers. In the world of computers, the 1950s came to be seen as the time when the "mainframe era" of computing began (see figure 2.1).

Because of their complexity and cost, these were housed in data centers that were air conditioned, had staffs that wrote software and designed applications (today often described as *algorithms*), ran the machines, and interacted with salesmen and repair personnel from the vendor providing the systems. These data centers were an extension of the previous tabulating departments, often also called the "IBM Room" or "IBM Department," after the vendor that supplied these earlier data processing machines.[12] By the mid-1960s, IBM dominated the mainframe market, with about 60 percent market share in the United States and normally around the world between 70 and 80 percent, depending on the country.[13] Data center staffs normally reported to the chief financial officer of the firm or agency because the preponderance of early processing of information consisted of accounting and inventory data.[14]

Users were aware of the limits of computers. A report written by GE employees in 1952 regarding the technology's functionality, reliability, and capacity limits concluded that "The computer is so much faster at routine computational work" and that its "inherent speed also makes it possible to test a variety of assumptions."[15] Already it was becoming obvious that computers could be used to process information when direct and indirect labor in handling data could be reduced and when "what if" analysis needed to

Figure 2.1 Early computer systems consisted of multiple specialized machines connected to each other to move data from one to another for storage, processing, and printing. In time, these became smaller and could do more work. Source: IBM Corporate Archives

be conducted. The central theme evident for the next four decades was the possibility of reducing the labor content of work.[16] As employees became more dependent on information to carry out their tasks, they increased the computer content of information work. Many tens of thousands of office and other clerical staff shrank in number as their tasks became automated or, if their employers were growing in size, did not proportionally increase in population.[17]

Historians assign the prize of being the first commercial use of a computer to GE's appliance manufacturing plant in Lexington, Kentucky. The company installed a system to do such work as inventory control, shop floor scheduling, and other manufacturing accounting. Key to this installation was that it received enormous publicity, thereby encouraging other firms to embrace computing.[18] By mid-decade when these systems were reliable enough to replace tabulators, economic benefits outstripped their high costs for hardware and staff in large enterprises and agencies. Not to be overlooked, the technology had evolved sufficiently to handle the variety and volume of information required. Improvements in performance, reliability, and capacity continued normally incrementally over the next seven decades. As part of that activity, the costs of processing information declined continuously, too, generating more productivity than any other measurable technology in the twentieth century, often at compounded rates of 20–22 percent *per year*.[19]

A half dozen applications became attractive in this first era of computing. Accounting built naturally on earlier data processing uses of tabulating equipment and so ported quickly over to these faster, higher-capacity new systems. Computers were integrated first into the calculation of financial reports, accounts payable, billing, accounts receivable, sales analysis, and inventory costs.[20] These calculations were further integrated with each other as a second phase in computing. Reports were produced more frequently and in more granular forms over time.[21] Industry specific applications began to appear, notably in banking and insurance, whose only inventory was largely data handled by hundreds or thousands of employees. Insurance companies were large enterprises with almost all their workers residing in offices or in sales.[22] In the 1950s and 1960s, banks standardized checks with magnetic ink character recognition, which made automated reading of these easier. Banks coordinated the transfer of inter-industry transactions increasingly using computers, such as in cashing checks from other banks.[23] In the insurance industries (health, life, and property), a long tradition existed of using all extant data processing technologies (adding machines, desktop calculators, tabulators) since the nineteenth century, so these firms were quick to embrace computers because they understood early why to do so.[24] Largely it was to reduce their extensive data entry and data processing activities which were frequent, subject to errors, and expensive to staff. The hunt was always on to improve data handling productivity.[25]

Manufacturing attracted the most attention of business managers, the public, and historians. It was the largest and most capital-intensive economic sector in most countries, certainly in the United States in the 1950s and 1960s. This industry embraced computing almost as quickly as banks and insurance firms. Applications varied more, however, and while building on earlier uses of data processing, they also added new ones. By the late 1950s, manufacturers were collecting and processing production data and shop floor information, conducting production forecasting, and automating shop floor activities (such as reporting on the use of parts and tasks completed). Manufacturers used computers to reduce the labor content of work, which was the basis of the business case for using computers and for gathering and using more information. Almost all manufacturing industries with large firms on both sides of the Atlantic Ocean used computers by the early 1960s.[26] Process industries, too, appropriated computers in the same way. In their case, they also needed information about the status of continuous flow operations, such as where oil was in a pipeline, using sensors as well since humans never saw the product, as it was always either in a pipe, at refineries, or in a machine.[27] Pulp and paper companies, petroleum and chemical plants used computers to process their accounting records and to model flows and schedule work.[28]

In addition to the intrinsic attractions of computerized information handling, almost every industry had inventory, so inventory information handling by computers became an attractive use of the technology, not just for manufacturers. By the mid-1960s, for example, the U.S. military services had more experience and larger deployments of inventory management applications than any private company. It seemed by then that "everyone" was automating inventory control practices in the United States, Western Europe, and Japan. Of all the early uses of computing, users were most often satisfied with the kinds of information computers spit out about inventories.[29]

COMPUTERS TALKING TO OTHER COMPUTERS EVERYWHERE, 1960–1980s

Then the world of computing changed. On April 7, 1964, one of the largest vendors of such equipment—IBM—introduced to the world a "family" of computers of varying sizes, including new input and output equipment, operating systems, and other software utilities, for a total of some 150 products. Nothing then or now came close to that sweep of introductions. Even today, when Apple introduces tablets or smartphones, maybe a half dozen products are brought out, not 150. The story of the IBM System 360 has gone down in history and business mythology as one of the most dramatic product and technology stories of the past century.[30] It was a development filled with high drama as much as it was a remarkable technological achievement.[31] From an information history perspective, its importance lay in the fact that if a data center was being asked to process more information than their installed computer system could handle—and those requests were increasingly made by the early 1960s—then a data processing manager's options were (a) not to do it, (b) add another expensive computer system with additional staff to do the work, or (c) rewrite all prior software applications to fit into a larger system. All three options were either unacceptable, too expensive, or fraught with technical risk.

System 360 held out the promise that IBM delivered in time to fix or eliminate those three ugly options: applications written for a smaller model 360 could be ported over to a larger machine in this family of computers with little or no modification (see figure 2.2). Second, computers could be swapped out over a weekend without necessarily having to replace input or output equipment. This was referred to as a "heart transplant." Third, with additional capacity, different collections of information could be integrated. For example, two name and address files, each in separate computers, say

Figure 2.2 This family of computers introduced the idea of compatibility among machines that could be added to or increased in the ability to handle more processing. As a result, computer usage across the industrialized world expanded rapidly in the 1960s and 1970s. *Source*: IBM Corporate Archives

used by billing and mail-order advertising, could now be combined into one list shared by both applications. That saved on the double cost of data entry and reduced the number of errors in such files.[32]

The demand for these and subsequent mainframe systems (meaning software, computers, and their peripheral equipment), once their functions were understood, grew almost exponentially over the next two decades. No longer could only large companies and agencies afford computing; soon mid-sized ones could too, even firms with as few as fifty employees. Systems became easier to use as more sophisticated programming languages were adopted, and in the 1960s commercially available software products that performed such standard calculations as accounting and inventory control.[33] These computer systems, and those that were subsequently introduced by IBM's competitors, were also more reliable and easier to use. Demand for all manner of computing caused the computer industry to expand at a compounded annual rate of over 20 percent during the second half of the 1960s.[34] For the first time, one could begin to think that "everyone" was using a computer, that society had now entered the "Information Age." This hubris was no longer just marketing buzz, but was now taken seriously by scholars, public officials, and seemingly the majority of middle and senior manager.[35]

The availability of this new generation of computing resulted in further reliance on digitized information. With such installations combining disk drives, terminals, and telecommunications, workers began accessing information online, as opposed to printing reports, or even waiting for one to be delivered. Because online processing spread during the 1960s and 1970s, by the 1980s the majority of information queries had transformed into the new mode. With the availability of PCs in the 1980s, too, workers became accustomed to looking up bits and pieces of information when needed, or to use a new term from the period, "on demand," and storing it on their own machines. Those capabilities encouraged increased access to even more diverse bodies of digitized information. By the end of the 1980s, hardly any office worker was immune from that practice in the industrialized world. E-mail began in the 1970s in a few large corporations and government agencies, but by the end of the 1980s was nearly ubiquitous in large public and private organizations.[36]

The U.S. Bureau of Labor Statistics surveyed all manner of industries in this period, tracking how workers and industries used computers, which is how we know, for the United States at least, so much about the diffusion and use of digitized information. Its studies demonstrated that the earliest and most extensive users of such information were in banking, electric and gas utilities, insurance, petroleum refining, printing and publishing, process industries, railroads, steel, and telephone communications, among others. A second wave of users that overlapped the first, but came later, included aerospace, air transportation, hosiery, intercity trucking, metalworking machinery producers and users, oil and gas extractors, and textile manufacturers. Laggards who came in a third wave included aluminum manufacturers, large bakeries, coal and copper mining, concrete manufacturers, metal fabricators, footwear, laundry services, forest products industries, meat processors, water transportation and wholesale trade, among others. As with the first and second wave adopters, the largest firms in each industry came first, followed by smaller ones as distributed computing and PCs brought down the cost of computing.[37]

Each industry tailored data to its needs. For example, the aerospace industry wrote software to model airplane designs and then numeric control machines that could be instructed to cut metal components. Apparel manufacturers learned how to instruct laser-cutting devices. Banks automated major functions and, in the late 1960s, added in the late 1960s ATM services, replete with all the necessary automated account record keeping required for such services. Hosiers learned to knit online, intercity trucking let computers handle the billing, accounting, and scheduling, while utilities relied on digital meter reading (satellites in the 1990s, too, to do some of this work). In some industries, a great deal of the information collected and used was turned over to machines and computers, such as in the petroleum industry to manage the extraction of oil, its transport to refineries, its processing, and then scheduling deliveries to customers. Other applications involved collecting, analyzing, and using data by machines with minimal or no human involvement, such as controlling the speed of railroad cars, switching, and traffic control. Point-of-sale systems in retail companies kept track of sales, determined what inventory was needed, even ordered it, studied patterns of consumption and purchase, and informed humans about all these activities on either a prearranged schedule or on demand. AT&T's systems automatically tested telephone lines, informing people only if there was something out of the ordinary; otherwise, the system took care of monitoring and making adjustments to line performance. By the end of the 1980s, automated navigational services had taken over many of the takeoff, in-flight activities, and landing tasks previously performed by pilots (although by law human pilots were required in the cockpit with the capability to override computers).[38]

By the end of the 1970s, new reasons joined earlier ones for using electrified information. In manufacturing alone, in addition to reducing expensive labor content in work, computing did work faster and better controlled its quality and products, which delighted leading information gurus such as W. Edwards Deming (1900–1993); and made it possible to change production specifications, schedules, and even products made on the shop floor, initiating an era of "flexible manufacturing."[39] One could now model and let computers optimize energy and raw materials consumption, which proved important in high-energy-use industries, such as steel and automotive manufacturing. Finally, complex processes became easier for humans to manage, such as in papermaking, where smooth flow of materials proved essential to avoid tearing rolls of paper while being made.

In short, users moved from digitizing preexisting collections of information in the 1950s and 1960s to doing new things that computerized information made possible in the 1960s and 1970s (often referred to as *augmentation*). Then they increasingly let computers do the work (called *automation*) in the 1970s and 1980s, telling people what they were doing if asked to, otherwise only reporting on exceptions to expectations (the "Houston, we have a problem" situation). Beginning in the 1960s, humans could begin going to computers to ask for advice (e.g., technical standards, rules, and company policies) for how to deal with situations, such as in customer relations, billing disputes, optimized designs for products (e.g., airplane wings and automated machine control systems), and to train new employees online. In other words, by the end of the 1970s, and even more extensively in the 1980s, machines were given the capability and authority to create, analyze, and use information in addition to human beings.[40]

During this period, computers came in smaller forms so that computing could also be distributed around an enterprise, all "talking" to each other and to the older mainframes, discussed further below. All of this happened essentially in the working lifespan of one career. By the 1980s, the ability of management to delegate authority to make decisions and to do work increasingly became possible, reducing even further the need for middle managers to collect and distribute information up and down the organization.

Then Came Personal Computers and Their Sequels, 1970–2000s

Deployment of PCs in companies and government agencies began in the late 1970s, but deployment seemed trivial when compared to what happened in the 1980s. After IBM and subsequently its rivals introduced functionally attractive and reasonably affordable PCs in the early 1980s, they took off, such that by the end of the decade, people were buying these by the millions per quarter.[41] Some companies were acquiring these literally by the truckload. By the early 1990s, it was not uncommon in large enterprises to see unopened newly arrived shipments of PCs sitting in closets and storerooms waiting to be distributed to employees.[42]

PCs were the first computers that individuals could afford to acquire and that were relatively easy to use; one did not have to be IT savvy to use these. The initial "killer apps" were word processors and spreadsheets, and in the second half of the decade, linking to telephone lines, thus, sharing information with someone elsewhere. Users began to tap into commercially available datasets and access internal ones at work. That communications innovation stimulated a massive new wave of adoptions, this time, too, because now e-mail became possible. By the early 1990s, tens of millions of machines were being sold every few months around the world.[43]

But corporate directors of computer operations had been complaining for some time about these little devices. In the late 1970s, they were telling IBM that it had to introduce a small desktop machine so that data processing managers could have a standard technology they could support; otherwise many incompatible systems would continue to appear, much as had occurred with mainframes in the 1950s and early 1960s that had proven so chaotic and unmanageable. IBM complied, and for several years everything seemed to be fine: engineers and marketing got their machines; others could use computing without having to fight their DP centers and financial managers for priority in providing new software tools. These were simple and could communicate, and they worked (see figure 2.3). But in the late 1980s, data center managers started grousing that the amount of computing power installed inside their companies and agencies, not under their control, equaled that of all their mainframes, with PC users now demanding access to corporate databases and "tech support" whenever they had issues with their desktop devices. What was going on? Just as bad for these managers, they were losing control over computing that they had enjoyed for their entire careers.

Figure 2.3 IBM's PC, introduced in 1981, led to a massive increase in the use of desktop computing, much as the System 360 had with mainframe computers in the 1960s. By the end of the 1980s, many other vendors were selling IBM-compatible PCs. *Source*: IBM Corporate Archives

When viewed through the lens of information, one could see profound changes underway that began before the arrival of the PC but then became turbocharged, such that nearly a half century later, they were still unfolding, much like the Big Bang of the Universe. Simply put, all during the first half of the twentieth century, institutional collections of data were nested in a few locations, such as in corporate and divisional headquarters and in data processing centers, as in those old "IBM Departments." Information went to a few localities where it was collected, analyzed, and reproduced, then distributed back out to all corners of these enterprises. Middle management had its purpose. As computers became easier to use and appeared in smaller sizes, beginning initially in the late 1960s with smaller mainframes, then more dramatically with "mini" computers that became popular with engineers and other smaller enterprises in the 1970s, digitized information began spreading out, diffusing to use the term preferred by students of the process. One could see this unrolling as engineering departments wanted to get out from under their perceived tyranny of financial and administrative managers who they still believed were only interested in accounting applications. Engineers saw that the technology was now making mathematical, engineering, and modeling uses inexpensive, fast, easy, and useful.

Other departments did the same, demanding that they be allowed to spend from their own budgets on these "Tools for Modern Times," to quote IBM's PC advertising tagline. They wanted, demanded, and received access to centralized digital files. People with minis and PCs began building their own libraries of digitized databases. When laptops came in the 1990s and tablets in the early 2000s, then smartphones, beginning in 2007, the process of information's diffusion repeated. Meanwhile, commercially available datasets became available, which could be used in tandem with internal databases, enriching information ecosystems of all departments, not simply offices.[44] New information was created even at the most outer points of a company, such as with Radio Frequency Identification (RFID) tags on grocery shelves to inform humans about inventory, extending even to barcodes on packages to inform about the sale of a box of cereal or to record that a package had been delivered to a warehouse (later to a home).[45] The list of examples of the diffusion of information to "the last mile" became endless, more so as electronic sensors were embedded in everything from man-made "things" to even being mounted on the backs of "murder hornets" to track their activities.

Large data centers no longer monopolized digitized information; they evolved into technical hubs that supported maintenance and operations of vast databases, networks, and data security. They increasingly supported individual end users encountering difficulties using their small computers. Knowledge about computing diffused too, to such an extent that by the end of the century it was not uncommon for the most senior executives to have personal experience managing implementation of large digitized information systems. They personally used e-mail, often with PCs instead of the old "dumb" green screen terminals of the 1970s and early 1980s. They understood the power of more and different types of information employees relied upon to do their work.

Meanwhile, each wave of new technology made it possible for people to create localized collections of information in digital formats equal to or greater in volume than what was possible, say, with a mainframe a decade or two earlier. That new activity resulted in the overall volume of digitized information everyone could work with growing massively. This happened to such an extent that today attempts to quantify how much or how fast information increased border on the absurd. One can cruise the Internet and find thousands of citations pointing to how many zettabytes of information exists, which is virtually meaningless as it is so much.[46] We address this issue in more detail in chapter four.[47] Such discussions had become almost irrelevant by the early 1990s, because the effectiveness of work and the success of individuals had become highly dependent on access and use of information.

To make all that new information useful, organizations began thinking in terms of processes that were information laden. It made possible complex operations. For example, how can one automobile factory assemble 300,000 vehicles in one year with 6,000 employees, with each car or truck made up of thousands of diverse parts from screws to chassis, rubber tires to metal frames and do all of this on time, within budget, profitably, regularly, and in one year? Information combined with disciplined work tasks was the answer.

Studying processes, W. Edwards Deming's mantra, made much of this operationally possible because processes atrophy (laws of physics teach that) and episodic events are disruptive too, so processes have to be constantly understood and modified. That could only happen with information.[48] By the end of the twentieth century, it was difficult, if not impossible, to find formalized processes in any industry not rich in information and the subject of nearly scientific-level disciplined study on a regular basis.

This became so much the case that devices used in many industries and applications increasingly also self-monitored performance, analyzed and adjusted operations with or without consulting human operators, beginning by the end of the 1980s. In the 1990s, new terms appeared to describe this behavior, for example, "analytics" and "business intelligence," which became subfields of statistics and managerial practices in their own right. Workers in front of a PC were now expected to use more numbers, statistics, quantitative analysis, modeling, forecasting, and predictions as normal work tasks.[49] These behaviors expanded as computing diffused throughout organizations but also evolved as process managers imposed analytics on top of traditional statistical data sets to provide modeling and "what if" alternative analysis to optimize decisions. Much of this was numeric but presented in graphical formats, crunched by statistical tools. Employees wanted answers; they did not want to have to conduct the mathematically tedious work involved. That they willingly delegated to digital tools.

As more data became available in more departments, and as employees became more accustomed to viewing their work practices as collections of processes, they took on what in hindsight one might refer to as a Lego view of the world: they began to connect together processes into mega-processes, held together with the glue of information. A widespread example of this process is what I came to think of as the "informed supply chain."[50] These are often the largest and most complex collections of processes in any organization. These were made possible by the combination of significant amounts of information and coordination. Every large enterprise operates supply chains, and every small organization is somehow a participant in someone else's.[51] Automotive manufacturers were some of the earliest enterprises to create formal supply chains. By the 1980s, an automotive factory connected to suppliers of all the components it needed to build a car. Each participant in the supply chain would be given information on what parts to make and when to deliver them to an automotive factory. Deliveries were assigned on a day-by-day basis as opposed to requesting delivery of large quantities of a product, say, monthly or quarterly. Increasingly, suppliers and factories shared the same databases, such as the production plan for cars or trucks, so that, for example, a tire supplier could plan the kinds and how many to deliver on what days to the factory and, in some instances, where on the shop floor.[52]

The reader undoubtedly has seen suppliers of potato chips, soft drinks, and bakery products do the same thing at grocery stores run by national chains, not grocery employees.

These supplies are global. In the case of automobile manufacturing, parts come from all over the world; with a few minor exceptions, all cars outside of China are manufactured with parts from many countries. A car made in Detroit is not an American car; it realistically is an international vehicle. While the COVID pandemic of 2020–2023 has started to result in the de-internationalization of some features of global supply chains, by the end of the 1980s, these had become the ubiquitous way of organizing manufacturing and logistics around the world, thus making them difficult to dismantle.[53]

In supply chains, every step is accompanied by information required to (a) receive instructions to perform a task, (b) do the work, (c) report that this was completed, and (d) generate metrics on such issues as how long it took, its quality, and that tasks were all passed off to the next step, and so forth. Built into the equipment, terminals used by participants, and sensors are software capabilities to create, move about, and analyze information as part of the process. All participants are kept up-to-date on their individual performance. Information is shared across departments within an enterprise and also with other companies, regardless of where they are located. The data crosses international boundaries and, again, usually regardless of local privacy and other data management regulations and laws. Legal, cultural, and technical issues related to information flows are resolved after a practice has been ongoing for a while.

The information infrastructures of supply chains work in most, if not all, industries. In 1913, when the United States established the Federal Reserve System (beginning operations in 1915), it created a de facto supply chain: a set of measurable processes for moving money from one bank to another, from the United States back and forth to international banking systems, with informational audit tracking consistent along the entire supply chain.[54] Now money moves electronically in all countries. Banks have their own for the same purposes. In the Internet era, they expanded further to include ever-smaller enterprises (e.g., through PayPal) and individuals (e.g., through Amazon.com).

Thinking in terms of a supply chain paradigm began in the mid-1960s when computers could be connected to each other using telephonic technology, Claude E. Shannon's old notion of electronic signals.[55] Managers were able to connect inventory management information to manufacturing and transportation data sets. By the end of the 1980s, these were called supply chains. Knowledge of best practices had spread widely, too.[56] These firms added other concurrently emerging sets of information, notably customer and market forecasts, to monitor demand for specific products. They began relying on modeling of potential new offerings and upon a growing collection of data about deliveries and costs that could be communicated to customer organizations by the mid-1970s, and to the individual customer in the early 2000s. Information became increasingly accessible in real time, that is to say, available first in days, then in hours and now often in minutes or seconds. Nothing seemed to stay in one place long. Parts inventories that used to sit in warehouses in the 1980s for close to two months dropped on average to less than forty days in the 1990s, to nearly one month in the early 2000s. That "just-in-time" approach worked well as long as wars and pandemics did not interfere; when such existential events did, the world discovered how fragile supply chains could be.[57]

The availability of information made it increasingly possible for vendors to modify their supply chains to anticipate customer demands—classic market survey work linked to supply chains—while the military did the same to inform its requirements based on models of new warfare tactics and strategies. These practices required refurbishing older versions of supply chains with increasing amounts of forecast data from the inception of an idea to the delivery of a new product or service. These changes in information evolved into an iterative process, a way of working that by the end of the century had been going on for so long that few could remember a time when that was not the case. Linked to all these activities were other collections of data to identify and anticipate challenges, breaks to supply chains, and modeling of potential solutions to problems.[58]

Finally, one should understand how information seeped into and came to characterize (indeed, dominate) supply chains. Until the 1980s, clerks collected information needed to operate supply chains, normally operating within the confines of one country or company. Then they began using spreadsheets, many in PCs and minis, but increasingly, too, in mainframes. These were used to document and calculate relationships between sets of data, such as parts and products, logistics, and product availabilities. The installation of barcodes and scanning made it possible to add more granular information down to individual components and products, and then, of course, to track their physical movement as these occurred. By the early 2000s, sensors had largely taken over much of the work of clerks in maintaining information about all aspects of a supply chain's activities. Supply chains had increasingly become automated, reducing the labor content of people while enhancing the role of rule-based data coming from sensors with software then communicating back out to them instructions on what to do next. Artificial intelligence (AI) was in full force by the early years of the twenty-first century. Meters, actuators, and GPS data continued to add to a supply chain's information ecosystem. Humans learned how to share information electronically, which proved essential to the value of supply chains.

Pattern recognition became the early new information that came out of these that could empower AI applications.[59]

While this may all sound like science fiction, much of it works as just described, but information, too, has its own problems. For one thing, not all firms were/are willing to share information across parts of someone else's supply chains. For another, since so much is automated (e.g., RFID signals, product planning, and forecasting), what information needs to be made visible to humans has become increasingly concerning, especially if someone is managing a global supply chain. It is easy to conjure up an image of a "war room" much like one pictures NASA having for launching rockets, with rows of large TV screens and people in short-sleeve shirts and headsets staring at them. But even at NASA, much information is automated, shielded from human eyes, which would otherwise have to be transmitted too slowly for human brains to process accurately. Ultimately, it all boils down to the problem librarians encountered in the 1800s: What is the right information that someone needs, and when?[60] Like those librarians, today's supply chain managers seek to organize work and information in ways that allow humans to accomplish their objectives.[61] While the COVID pandemic and growing international tensions are causing many companies to consolidate their supply chains within the confines of a nation's border, in effect de-internalizing these, the fundamentals of the role of information in these applications remain the same.

Tour the information ecosystem of a modern corporation or government agency and one finds information everywhere. In one study of that proposition, new product development and marketing seemed to have almost as rich a body of information as supply chains. All of this information was additional to accounting, financial, and inventory data that historically had been core components of a corporation's information ecosystem for 150 years.[62]

Digital Plumbing and Information Infrastructures

In the early 2000s, it was becoming clear that corporations and large government agencies had incrementally and almost silently, behind the veil of internal operations, created a network of digital plumbing through which information flowed. We can think of it as plumbing because, as with the movement of water inside a building, pipes behind the walls moved it about without being seen (or even heard), propelled by end users turning on faucets or using appliances. It was similar with information. Digital plumbing was also an infrastructure that made it easier to rely on information in increasingly varied forms over the past seven decades, thereby contributing to the changing nature of work. Information was available on demand by just accessing a PC or digital file. The plumbing analogy returns us to the argument made in the first sentence of this chapter: that the combination of using all manner of information technologies and information fundamentally changed work. As with real pipes and periodic changes in bathroom and kitchen fixtures, infrastructure changes. The same happens with digital plumbing. It does so today, when, for example, one goes to work in a different company, only to find that employees no longer store digital data in Google files, but rather in Slack, or that Instagram is "out," while something new is "in." The point is that in either place, an information infrastructure exists and becomes how one interacts with information.[63]

It is an old story, of course. At IBM, for example, an early e-mail system (PROFS) was replaced with Lotus Notes in the 1990s, and in the 2010s with yet additional different communications tools. Roughly 4 percent of all office or knowledge workers worked out of their homes using the Internet to transmit messages and deliverables by 2010, but with the pandemic beginning in 2020, that population increased by an estimated one order of magnitude within weeks.[64] Crowded offices with cubicles were "out," social distancing with one-way paths for walking in an office or factory was "in." While such changes in work practices and facilities unfolded, the capabilities of combined IT and information at work made much of this possible. We had reached the point where discussions about information were also digital conversations, just as discourse about information in the nineteenth century perforce was shaped by the architecture of "the book." By the time the pandemic was over, so many employees had become used to working at home through their extended digital plumbing that they balked at returning to offices. As a result, many office buildings are largely empty in such cities as New York, San Francisco, and numerous cities in Europe with serious consequences for banks holding commercial loans and cities losing tax revenues.

It was not uncommon for the world's industrial economies to spend 6–7 percent of their GDP on information by the end of the twentieth century. Almost every office and factory in moderate-to-advanced economies was wired for electricity, coaxial cable, or fiber. On either side of the Atlantic Ocean, people had wall outlets into which they could plug their terminals or laptops. The wall outlets worked, their laptops could handle American and European electricity, the Internet was everywhere, Gmail accessible essentially around the world, and Netflix was there to entertain.

The U.S. Department of Commerce, U.S. Bureau of Economic Analysis, and equivalent agencies around the world informed us about how much societies were spending on telecommunications. If you worked in a building anytime after roughly 1975 and witnessed any renovations, people were "pulling cables" through the walls; meanwhile, data centers sprouted like weeds, causing mayors in large cities to complain that communities might run out of electricity. Organizations were building physical and information networks, making it physically and operationally possible to move information to almost any place and increasingly with individuals. The Internet had become big business, no longer a government communications network.[65]

Careers were made. Lower-level data processing managers of the 1960s became middle managers in the 1970s, and in the 1980s a divisional vice president of Management Information Systems (MIS)—an executive. It did not end there. In the 1990s, he, and increasingly she, became a corporate Chief Information Officer (CIO), paid and appraised increasingly like senior executives. Information was now a "corporate asset" to be protected, of course, but just as important to be made as ubiquitous as air conditioning. Expenditures on IT had moved from a range of 6–7 percent of a corporate or agency budget to averages closer to 5 to 9 percent in the 1990s.[66]

By the early 2000s, when I began to think of digital plumbing as ubiquitous as running water, CIOs were explaining to me that their digital infrastructures had become "systems of systems" overlaid on each other and increasingly integrated. Their organizations had become collections of processes and information ecosystems; they no longer thought of buildings full of people. All processes had information, and all information had processes. Both facets were documented, tracked, and understood. City managers, for instance, saw their communities as collections of systems supporting health, transportation, fire protection, personal security, traffic management, water supplies, education, and economic development.[67] Employees were embedded in these systems. Corporations saw themselves the same way. There was less talk about information, hardware, and software, more about information, processes, and ecosystems. Data security, project management, and monitoring of information took up growing percentages of one's workload.[68]

The information ephemera of the 1990s included e-mail, ever-longer PowerPoint slide decks, graphics everywhere, and, of course databases and spreadsheets. All of these existed before Google. The Internet pushed the digital plumbing to a new level of density. In the 1950s through the 1970s, there existed what then were called "islands of automation," places in an enterprise with much computing, while the rest a relative information desert. By the 1990s, these "islands of automation" had disappeared, replaced with new ones that were relative. That is to say, while "everyone" had access to digitized information, some had less, creating information deserts, offices that in earlier times might have uncharitably been called "backward," but now because they lacked the latest IT, or its employees were not as computer literate as expected. That was when jokes circulated about how ten-year-old boys knew more about computers than their parents.[69] The humor worked because all good humor is a fractured look at reality; the 22-year-old employee who relied on the digital plumbing far more than her 50-year-old manager. Islands of automation could now be thought of as continents of automation.[70]

One important effect of digital plumbing on work concerned who made the decisions and where they were made. We have discussed this development elsewhere, but it needs to be linked to digital plumbing. Supply chains spread work horizontally, often around the world within an organization. Deming's insistence on viewing work as collections of measurable processes encouraged management to move decision-making downward in the hierarchy of their organizations to where the activity occurred, following the old British idea of doing things at the coalface.[71] Technology made it possible to situate information wherever it was needed in the forms and required by each class of user.

CEOs no longer could order people to do big things, even to implement strategic decisions. Too many people were buzzing around in processes using information. To coordinate and integrate, CEOs had to state intentions, persuade employees to pursue desired directions, and then hope they took actions to fulfill those wishes. Like mayors and politicians, they had to post advertisements at airport baggage collection points to communicate corporate intentions; mass e-mails did too; later, advertisements and messages aimed at their own employees through Facebook, LinkedIn, and Google. A senior executive of the 1970s would have found this a strange world.[72]

We have already mentioned that software increasingly received the authority to collect information and take actions. These "expert systems" evolved over a forty-year period by the time computer scientists declared the emergence of a renaissance in AI in the early 2000s. When IBM, Amazon, and Microsoft announced that their central offerings would be cloud and AI after 2010, these were no longer revolutionary thoughts, but rather further evolution in a nearly 50-year-old digitized information ecosystem, the digital plumbing of the late twentieth century. So machines (largely intelligent sensors), software, and people were continuing to add and change information.

One additional transformation in how information affected work can be thought of as convergence. This is the act of linking together technologies and information, not simply concurrent co-dependent needs of one to the other. Rather, it is a silent, nearly invisible activity, such as phone messages left on one's landline in a pre-2010 office telephone also posted concurrently to one's e-mail system and, therefore, deposited in a computer (today a server) hosting one's digital plumbing. By the 1990s, one could be on a conference call, stare at slide presentations on a screen, and make running comments using texting on either their landline or laptop/PC; in the 2000s, the same could take place during webinars.

Digital plumbing has made it possible for people to (a) collaborate while physically separated down the hall or around the world in any time zone; (b) use and manage vast bodies of data, not possible even in the late 1970s by comparison, including data collated from multiple sources in nearly real time, creating expectations that all of one's colleagues would have access to similarly large collections of information; and (c) workers—both blue and white collar—have had to get used to working at all hours of the day or night, including on weekends and local holidays. Some readers undoubtedly have had the experience of a Monday morning staff meeting in Australia translating into a horribly early Sunday meeting in the United States or Europe. As late as the 1970s, such working conditions would have been seen as science fiction, with only CEOs subject to such working conditions.

Welcome to the Internet, 2000–2010s

By the time the Internet became integral to the digital plumbing in all phases of people's lives, much had been digitized, especially work converted to new forms that made appropriation of the Internet less radical than was once thought to be the case. With a quarter-century of experience with wide access to the Internet now as part of our collective experience, one can see that the Internet added more content in more diverse ways across a broader spectrum of topics in ways more accessible than before. But its pattern of use had been shaped before the 1990s.

Our understanding of how the Internet came about and spread so quickly is increasingly growing, but we resist delving into its history to focus on information. However, several historical circumstances help to explain why it was possible for people from all walks of life to seize upon that technology—network—and to use it quickly to shape their work and private information ecosystems. In the process of doing that, billions of people shaped the nature and type of information available to them and, today, to more machines and sensors than there are humans. Today, sensors and servers use the Internet more than people; Chinese is the most widely used language on the Internet; and this technology has been in use for over a half century![73] Yes, you read that correctly—a half century—which is adequate time for a technology and its uses to shape the nature, type of information, and quantity available to more than just human beings.

The Internet was a byproduct of the American government's interest in developing information and communications technologies and networks, along with weapons as part of its Cold War defense strategy.[74] By the early 1970s, what came many years later to be called the Internet had been developed. Its essential feature was the ability to send packets of information through a network rather than a continuous stream of data as in standard telephone lines. Additionally, the Internet was designed for these packets of data (Shannon's idea of signals) to course through alternative paths should one not be functioning, rather than the old way of going from point A to point B. The combination of breaking up data into packets and giving them the ability to take alternative paths to their final destination was purposeful to increase the ability of a communications network to continue operating in the event of a nuclear war. In the 1970s and 1980s, it worked and defense contractors, and increasingly individuals and academic, military, and defense organizations used the Internet. But it was, to be kind, "techy" e-mail and file transfers, despite incremental improvements by computer scientists improving its operations.[75]

A breakthrough came in the early 1990s with the development of the World Wide Web, which, to keep our narrative simple, made it possible to write commands in human languages—in the beginning, English—rather than in combinations of characters and symbols that felt to many as if they were communicating in some ancient language. When one could write and read real words, attach data files, and later work with images, the Internet reached a technological tipping point where literally billions of people could use it. Between the mid-1990s and the end of the century, its use grew exponentially. Non-technical people migrated to it, and corporations and government agencies began to use it to distribute information, then to communicate within their institutions and externally to other organizations. New businesses sprouted to sell connections and information and for corporations to communicate with customers. E-mail became a "killer app," while organizations began loading information about their products and services onto the Internet. By the early 2000s, it had become de rigueur to have a website and to make it possible for individuals to communicate with enterprises and agencies.[76]

The newsworthiness of the Internet's story in the twenty-first century is about diffusion. It went from possibly as many as 45 million users in 1996 to over half of humanity barely over two decades later, accessible around the world.[77] When smartphone usage took off worldwide after 2007, it quickly became the single most used device for accessing the Internet and its continuously increasing amount of information. Why? Because these phones were inexpensive when compared to PCs and terminals, easy to use, portable, and nations quickly were able to create Internet access facilities.[78]

There are, however, additional factors to consider yet to be studied by historians and others, although partially already done by economists. The first is the most obvious: that individuals and organizations in the most prosperous economies could afford the price of admission: the cost of a PC, which between the 1990s and early 2000s dropped from several thousand dollars to $500. One needed to use these devices to access the "Information Highway," just as a telegraph operator in the 1800s needed a "telegraph" to transmit signals down a line. So they acquired the new technologies.[79]

A more consequential second factor was satisfaction of a precondition: people had already invested financially and cognitively in the notion of creating, transmitting, receiving, and using information electronically.[80] Global annual sales of PCs increased from 700,000 in 1980 to 260 million in 2012.[81] Another investment—smartphones—rose from 23,000 portable phones in 1980 to smartphone sales in 2015 reaching 7.2 billion.[82] That latter number equates to one for every man, woman, child, and baby in the world. Microsoft claimed that 1.2 billion people in 2016 were using its Office products.[83] Amazon stored tens of trillions of digital "objects" by 2013.[84] A reliable report estimated that 35.6 million residents in the United States used voice-activated "smart speakers" in 2017, such as Echo.[85] Adoption statistics from the 1980s to the present documented that worldwide incremental diffusion of digital devices occurred from the 1980s to the mid-1990s, but soared after the arrival of the World Wide Web. By the end of the decade, trend lines spiraled upward in an almost hockey stick fashion, with mobile devices superseding PCs in 2012 when the smaller units were less expensive but performed essentially the same tasks.[86]

Those are the most obvious data points. But behind these is a less evident one. People had been acquiring digital devices since the consumer availability of microwave ovens in the late 1960s, with their dramatic takeoff in the 1970s. As new digital devices appeared, people incrementally learned how to judge when and why to acquire them. Think in terms of consumers behaving as if they had internalized a Moore's Law mentality, which in practice meant that with each new technology, consumers were able to make a decision on whether to acquire it or not faster than the previous time.[87] They had learned not to acquire these too early in the life of the technology because these would decline in cost and increase in function by a roughly 20 percent compounded rate each year.[88] They did not know Moore's Law, but sales patterns for each new generation of technology demonstrated the behavior, not simply that the cost of information handling devices dropped in price.

Consumers went from microwave ovens to digital watches, to PCs, next to laptops, flip phones, to digital cameras, then to a mix of tablets, iPods, and on to smartphones, and most recently to digital doorbells and thermometers. Statistics on when 25, 50, or 75 percent of the public had acquired any of these information platforms shortened from the time of introduction to when they achieved these levels of appropriation (see tables 2.3 and 2.4). It is a pattern of adoption awaiting the fine hand of an economist or historian to explain more fully, but it is the pattern of accumulating digitized information over the course of a half century that helps explain how people changed information. They shifted from reliance on information in paper formats to electrical ones.

By the 2010s, it seemed few people wanted to share paper copies of anything. Folks asked that you send them digital ones: scanned documents and data files. Even traditionally paper-oriented documents went digital. Being able to sign a contract online became a cryptographic exercise, enabling someone to sign it and transmit their signature—a data file—to another party

Table 2.3 Years from Introduction to over 75 Percent Adoption, Sample Digital Products*

Device	Years to 75% Adoption		Years to Estimated 25% Adoption
Microwave oven	1967–1992	(15 years)	Unknown
Digital watch	1973–1985	(12 years)	Unknown
PC	1978–2002	(24 years)	16 years
Portable phone	1978–2003	(25 years)	13 years
VCR	1988–1993	(5 years)	3 years
Internet	1993–2015	(23 years)	7 years
Digital camera	1990–	Never reached	6 years
Smart phone	2007–2017	(10 years)	3 years
Flat screens	1997–2007	(10 years)	3–4 years

Source: Census Bureau, U.S. Department of Commerce.
*Dates and percentages are estimates based on multiple chronologies and statistical data compiled using different data and calculating methods. Digital cameras were replaced for many people by smartphone built-in cameras after 2007.

Table 2.4 Adoption Rates by U.S. Homes of Major Digital Technologies Measured by Years

Device	Years to 50% Adoption by Homes
PCs	19 years
Cell phones	14 years
VCRs	12 years
CD players	11 years
Internet access	10 years
Digital TVs	10 years
DVD players	7 years
MP3 players	6 years

Source: Adapted from U.S. government sources by Adam Thierer, "On Measuring Technology Diffusion Rates," *Technology Liberation Front,* May 28, 2009, https://techliberation.com/2009/05/28/on-measuring-technology-diffusion-rates/ (Accessed May 9, 2023).

with both parties confident of its legitimacy.[89] It is now a common practice in the United States, across the European Union, parts of Africa, India, Brazil, Mexico, and various other Latin American and Middle Eastern countries.[90] Conceived in 1976, it did not take off until the 1990s. In the United States, as in many other countries, laws governing contracts had to be modified to make these legally binding. Concerns regarding security on the part of the technologists, lawyers, legislators, citizens, and business professionals were sufficiently satisfied, and so became an efficient norm by the turn of the century. In the United States, for example, the U.S. Congress passed enabling legislation in 2000, known as the "Electronic Signatures in Global and National Commerce Act," or ESIGN Act.[91]

The distinguished Washington D.C.-based Brookings Institution published a study on the degree of digitization of American work in late 2017, ranking professions. Explaining its methodology need not detain us here, rather some key findings.[92] The most obvious is that the greater the amount of organized information one used in their profession, the higher the score. Software developers, computer systems analysts, financial managers, and lawyers ranked the highest. But right below them came automotive mechanics, nurses, and office clerks. Yet, even the least digitized jobs relied on computing information as well: security guards, cooks, construction laborers, and personal care aides.[93] A close examination of job descriptions provided by the U.S. Department of Labor published over the past 150 years clearly documents the increased reliance on information across hundreds of professions and, for our purposes, in digital forms since the 1980s.[94] In short, employers and tens of millions of workers continued to transform information, with the greatest rise in their use of digitized versions occurring after the wide availability of the Internet. But remember, they had a heritage of being exposed to electrified information predating the World Wide Web for nearly two decades.

The Brooking's study offered a peak at the future worth summarizing. First, "The U.S. economy is digitizing at an extremely rapid pace." Nearly 100 percent of all workers either held "highly" or "moderately" digital jobs.[95] Second, "The degree and pace of change of digitization vary widely across occupations and industries." Between 2002 and 2016, the digital content of 517 out of 545 occupations (representing 90 percent of the U.S. workforce) rose.[96] Third, "Digitization is associated with increased pay and job resiliency in the face of automation."[97] Fourth, "The extent of digitization varies widely across places and is strongly associated with variations in regional economic performance." As economic standards of living improved and work became more information intense, the availability of information in digital forms increased, and facts became more evenly available. When that was not the case, such circumstances could largely be attributed to *relatively* less investment in regional communications infrastructure. This circumstance is normally referred to as a "digital divide," a reality sadly made abundant during the pandemic in the 2020s when children in poorer parts of a nation had less access to online schooling than those in economically more prosperous parts of the country, including in the United States.[98]

CONCLUSIONS

This is an opportune time to sum up how work changed as a result of everyone, from a new employee to the most senior leader of an organization awash in information by the start of the third decade of the twenty-first century. Some generalizations and conclusions can be drawn regardless of occupation or circumstance: academic, business employee, public official, teacher, retailer, even the unemployed or children. That is a remarkable statement to make, but this circumstance had been building for well over a century, picking up speed especially after the arrival of PCs, before accelerating even more when the Internet filled quickly with vast quantities of content and an unimaginable number of users, both human and otherwise. But several patterns of behavior—call them consequences of much new data—became evident.

First, and largely today thanks to smartphones, the bulk of employees, those who were not employed, and a large percentage of children over the age of fourteen in over half the world have nearly instant access to much information. Questions could be answered "on demand" to settle a bar bet, to get instructions on how to do something, or to make a fact-based decision. Technology platforms were no longer barriers; one could access a corporation's databases used at work through a smartphone in the forest as well as from a PC in an office. The convenience of access encouraged greater reliance on information; we know that because Google, the most widely used search engine outside of China, monitors this activity, as do other traffic monitors.[99]

Second, the information we now have relevant to work activities is organized into databases by topic and in relational databases, which allow one to tap into other sources of information and by process. This information is updated in many instances in real time or certainly frequently. For example, major news outlets used to update a current story perhaps once a day or every two or three days; now one normally sees updates several times a day, often with the source telling the reader how many minutes ago a story was updated. Sensors, software, and clocks made all that possible, but in the process created an appetite for very current information. A century ago, one-week- or two-week-old news was fine; during World War II, people saw weekly news stories at movie theaters on Saturday afternoons. When Abraham Lincoln was assassinated in 1865, it took days for the news to get around; when John F. Kennedy died in 1963, most Americans learned about it in anywhere from 15 minutes to two hours. Today, when a famous musician dies, the Internet is abuzz with comments and obituaries in less than 5 minutes. So, people think in terms of databases, processes, "just-in-time," and "on demand."

Third, simulations and models coupled with numeric data and options (scenarios) became continuously more attractive as computing made it possible to create more and varied narratives. Many of these turned out to be true enough to influence behavior and decisions. For example, in 2020–2021 during the COVID-19 pandemic, government officials in many countries ran models to describe potential scenarios for what would happen if they took one action or another, defining how many people would become ill, and how many would likely die. Until any of these outcomes occurred, it was all supposition, but the quality of these models became so highly respected that they were treated as facts, as accurate forecasts even though technically these were "fictions." Doctors, scientists, presidents, economists, CEOs, reporters, and citizens fixated on these. Information came in attractive multiple forms in all disciplines and industries: charts, graphs, and tables, moving charts that displayed changes over time, such as demographic data.[100] Much of it came in combinations of different formats, not just only as tables or graphs. It became nearly impossible at mid-manager levels to plan or make decisions without relying on models and scenarios almost as much as on facts of events that had already occurred, such as last week's sales statistics, or on one's experiences.

Fourth, because information could be moved about much more easily and inexpensively in ever-larger amounts and frequencies after computers could communicate with each other, people did that. Recall the example of the automotive supply chain that transcends companies and continents. When applying for a grant, such as the Information Technology History Society might do, it could ship a copy of its financials to a foundation in seconds, demonstrating to a potential donor exactly how it spends its money and how its activities would change if funded. An applicant for a position loads a copy of their resume into an employment agency's database and can modify its content to emphasize in a more tailored manner a position they are seeking quickly, often in minutes. In short, information could be modified with considerable ease. As a result, enterprises and public agencies morphed into information ecologies, or more precisely into information ecosystems. That is to say, more collections of facts than of buildings and inventories. In the process, information became more tailored to ever-narrower audiences. By 2010, one could speak of a *digital style* of working: data-intensive, mobile, reliant on models, and accessible.[101]

The Balkanization of information became a byproduct of so many facts, reinforcing tribes we call academic disciplines, and giving individuals confidence that they had a "handle on things" due to their access to varying qualities and types of information. Harry Collins, a British social scientist, attributes many causes to this development but does not lay one at the feet of just technologies. New ideas and socially constructed communities (i.e., academic disciplines) contributed to such developments. He describes a consequence: "when it comes to professional expertise, it is not what you know but what others think you know that counts."[102]

In this chapter, we used a wide aperture to see information as it appeared in work in general. Because that broadening of our perspective did not offer a complete vista of the nature and role of information over the past half century, we need to ask: To what extent do we live in an Information Age? Or to be elementary: Do we live in an Information Age? Answering such questions is the task taken up in the next chapter.

Chapter 3

Do We Live in an Information Age?

But any knowledge at all presupposes a world view, and the problem about sharing information is that where world views are in conflict, there will be little agreement about what kinds of knowledge are relevant and valid.

—Arnold Pacey, 1983[1]

We now live, we are told, in an information society and this is considered to be a very important development. But what does it mean? And what are the consequences?

—Michael Buckland, 2017[2]

If the title of this chapter looks familiar, it is because many hundreds of commentators in multiple disciplines have opined on the question. A second reason is my fault, because I have rewritten and published iterations of this chapter at least twice before the version presented here. My excuse is that every time I explored this question, the visibility of information in societies around the world had increased from before. That, in turn, was the result of new information activities that made more evident in some different way the role of facts, for many, the centrality of it in their lives. In this iteration, we have experienced a massive increase in the role of "fake news" in American and global politics, now resulting in worldwide discussions about the role of economic and political surveillance, and even more mysterious but very relevant, that of AI. As of this writing (2024), the U.S. Congress is contemplating legislation to restrict the activities of Facebook and Google, while many see Amazon as knowing too much about us individually; the European Union is already ahead on this front, while China and Russia implanted controls years ago; the latter more what one could express and facilities to gain greater access to one's private information. All of these developments are compounded by the practice of commentators throwing around the phrase "Information Age" without paying much attention to how they do that. To be blunt about it, they are sloppy while historians treat the issue almost with the rigor of a scientific discipline. I side with the historians, my home discipline. But before jumping into today's iteration, historical perspectives can be useful by going back a half century.

The 1970s have not been seen as tumultuous as, say, the 1960s, but they were still turbulent. The Vietnam War of the 1960s continued for a half decade into the 1970s. The world stepped off the gold standard, leading to inflation and other financial churn. The West experienced two severe rounds of oil shortages, resulting in long lines at gas stations on both sides of the Atlantic. Boomers still crowded universities to near overcapacity in scores of countries. The American Civil Rights movement had yet to calm, while the Watergate crisis challenged the public's trust in government and other institutions. The "Computer Revolution" operated at full throttle inside corporations and government agencies, but before PCs began homesteading in homes, offices, and small businesses. The Beatles were in full swing. American folk music still rocked. It was not a time when one would expect entire nations to contemplate what kind of society they were creating, except for a few philosophers, of course, a small number of novelists, and a handful of academics sprinkled across the United States and Western Europe. But that is exactly what happened. Why?

All societies nurture self-images that inform how individuals should behave, children should be raised, and institutions should function. Cultural, social, religious, economic, and political disputes—and the 1960s and 1970s were iconic periods replete with such churn—are normal subjects of interest. While today many would consider themselves as living in a world flooded with information, one would not think that was the case in the 1960s or the 1970s. But as this chapter demonstrates, that perspective had already begun to seep into the worldviews of people in the richest economies. People began to think about what a world filled with information and computers might look like, and even how to define the criteria for determining what forms it could take. For some, perhaps many, social orders were viewed as disconnected from prior eras, offering instead a future shock to the social fabric of a nation. The previous chapter explained how a world filled with information had expanded and shaped the nature of work and management within corporations, public institutions, and government agencies. After the 1970s, subdued notions about the role of information became unmoored as computing appeared in all manner of spaces, even reaching into our pockets and purses in the form of smartphones by the 2010s, and for many managers and elite knowledge workers earlier into their ears in the form of Bluetooth communications. In the process, information seemed in ascendancy as a feature of modern society, making it central to modern life and obvious to many observers.

This chapter reports on dialogues held within institutional and professional circles that burst out into the larger world of whole nations. Because that happened, we need to ask the question that serves as this chapter's title. As computing invaded homes (PCs), later nestled into one's luggage and briefcases in the form of laptops, and finally morphed into their hands as smartphones, it became obvious that everyone was now living in *the* Information Age. It was inevitable; one had to adjust to that reality so as to get on with life. That certainty about humanity's relations with computing was now seen as a permanent reality, an acceptance reinforced by the presence of computing for over half a century by the early 2000s.[3] But was that really so?

As I began writing this chapter, the world was subsumed in the COVID-19 pandemic. All the news and most blogs and Facebook postings focused on the crisis facing humanity. Cats, baby pictures, and recipes no longer dominated our interests, although they never quite went away. It seemed that many issues were chained to this medical crisis. Almost all economic and political perspectives and actions filtered through the harsh biological realities imposed on all. Soon after, too, Russia's invasion of Ukraine and another round of elections in the United States and Europe were all replete with social media interest, including fake news. Yet, South Korea was successfully implementing "contact tracing," using one's smartphone and GPS, national governments wanted communications companies and Internet service providers to do the same, while manufacturers of digital thermometers were reporting "hot spots" of increased human temperatures as indicators of potential next outbreaks. Proposals for tapping digital information were suggested in the United States and Western Europe, which clashed with societal values regarding privacy and fears of "big brother" government using such access in potentially sinister ways. Big data and massively increased amounts of information continued to loom in our lives; yet there was less discussion about whether we lived in an Information Age. Rather, it was more that we now occupied some sort of Pandemic Age, leaving behind an earlier one characterized by economic globalization and all those international supply chains.

But the volume of information that existed before the pandemic had not shrunk; in fact, traffic on the Internet increased, as people remained sequestered in their homes.[4] European governments asked Netflix to slow its transmissions so as to not clog the Internet; Google saw searches rise month over month. Knowledge workers sat in front of their screens Zooming or typing. If humanity ever lived in an Information Age, that time was now. It was also when information about the biological and medical realities of COVID-19 was insufficiently understood. It became a time when our confidence in knowing how things could be managed was challenged by unseen microbes ignoring known facts as they infected and killed people almost randomly, leaving survivors unable to control their mortality. Too many millions of people did not trust the barrage of information aimed at them to get Covid vaccine shots. Anti-vaccinators had weaponized medical information and their strategy was working.

In this chapter, I explore the earlier discussion about living in an information age, summarize the cases for and against that idea, and identify what effects such discussions have on how people view and use information. This chapter begins with insights historians culled about the nature and practice of naming an age, because titles matter. Names affect how people see each other. "The Greatest Generation" showered enormous respect on any adult who lived during World War II, while recently, being a "Boomer" conjured up perceptions of a selfish generation, one described as rebellious, that worked hard, but felt too entitled.

THE USEFULNESS OF A NAME AND HOW HISTORIANS CREATE THEM

Endowing a period with a name helps to give it identity, to shape what we think are its salient features. This allows an historian to signal that he or she is viewing a topic as political or economic history, for example. Names can also be changed. The Victorian Era became known later, too, as the Second Industrial Revolution. The Nuclear Age transformed into the Cold War. Media and economic pundits have used the monikers "The Information Age" and the "Digital Economy" for decades. The further back in time one goes to label an era, the easier it is for the nickname to stick as consensus or habit builds in various academic and media communities around how best to characterize it. The "Age of Jackson" of the 1820–1840s rightfully declares that American politics had transformed sharply from prior norms. The "New Deal" of the 1930s signaled a dramatic break from the politics and policies of the 1920s of President Herbert Hoover's failed response to the "Great Depression." "Camelot" is barely surviving as the nickname for the presidency of John F. Kennedy in the early 1960s. Some historians name an era based on artifacts, such as "The Bronze Age," while art historians think in terms of periods dominated by styles of art, such as Impressionism or Modernism. Political scientists were not shy either about putting labels on periods: "Fascism," "Democracy," "Communism," or like their historian cohorts, title a period as the "Russian Revolution" (which could mean anything for Russia from the 1870s to the 1930s, or just 1917), or the nebulous, "Modern World."[5]

It is thus a behavior with limitations, such as one sees when a child is given a name that sends mixed messages. There is the famous American country-western Johnny Cash song, "A Boy Named Sue" (1969) in which he sings a story of the difficulties encountered by a boy raised with a girl's name. "Pat" is both a girl's and a boy's name in the United States. It has been the butt of many jokes about the gender identification of its owner, made famous by a TV comedy show, *Saturday Night Live*.[6] The biggest limitation in naming is that a single feature is emphasized about a period, thereby deemphasizing other characteristics that are either just as important or, depending on one's perspective, possibly more so. Calling a time the Bronze Age minimizes the fact that its residents also used stone and iron and began building towns. Are those practices not potentially worthy of calling out through a name? Archeologists are sticking with their identifier.

What about the problem of chronology, how does one resolve that issue? Historians are not in full agreement on when the Bronze Age began or ended, nor when to end the Jacksonian Era, or for that matter, specific events. The Great War became quickly known as World War I (1914–1918 in Europe while in the United States, 1917–1918) followed by World War II. By the end of the century, historians had started to think of those two world wars as essentially one with pauses in between, hence now trying out new names, such as the "Second Thirty Years' War," "European Civil War," and the "Long War," among others.[7]

A third limitation is that one could impose on an earlier era a name influenced by current views or values, thereby emphasizing some issues more important today than in a prior time. Naming an age purposefully with that in mind is fine if the historian explains that, for example, she is interested in only emphasizing the role of women, slaves, or children.[8] Historians looking at the role of information in American society struggled with what to call the period, wanting to emphasize that information was a major force used by residents of the Second Industrial Revolution to shape the economy and culture of the United States in either recent times or as far back as the seventeenth century. An anonymous reviewer for *Birth of Modern Facts* wanted to know why I did not discuss the Third Industrial Revolution, while a week before writing this paragraph a commentator on television mentioned features of the Fourth Industrial Revolution. Historians are not quite ready to designate a "Third" or "Fourth" revolution. In this instance, they have even resisted labeling all these periods as an aggregate Information Age.[9] But an egregious example of imposing today's views on the past came from a journalist: *The Victorian Internet: The Remarkable Story of the Telegraph and the Nineteenth Century's On-Line Pioneers.*[10] The telegraph was not the predecessor to the Internet, nor is the concept of either online or the Internet largely comparable. Each was unique unto itself. The first was invented in the 1830–1840s, the second over a century later.

Historians frequently encounter these problems and are facing them again with respect to information, computing, and globalization, among others. But they have learned a few tricks about how to deal with such naming dilemmas. Normally, they look through a specific lens to establish discernible chronologies: economic, political, and military events are obvious examples. When wars begin and end are too, or 1870 for roughly the takeoff of the Second Industrial Revolution, even though they know that prior to a specific chronology one could add events, such as manufacturing in New England in the 1840s, or Chinese-Japanese conflicts in Asia and the internationalization of the Spanish Civil War in 1936 before World War II as part of the larger conflict's birth date. Regarding an information age, one might begin by asking when computers came into wide use, or were in use. Wide use, which would be subject to questions about what does wide use mean can be punted by saying when commercial computers came into initial appropriation, which means 1950–1952, leaving aside intramural historical debates about "firsts." But, even here what are we to say about punch-card tabulators, which preceded computers in large organizations by a half century, adding machines by 80 years, and so forth? So, historians begin by talking about framing through such lenses as chronology or theme.

If anything should be obvious from discussions in the previous chapter, it is that technological innovations—and there were a great many of them in data processing—should only be one of several considerations. No doubt mainframes, PCs, and the Internet are important milestones not only because of the functions they made available, but because historians should also be expected to look for massive uplifts in utilization. Moving from a handful of mainframes in the early 1950s to thousands by the end of the decade is an example; the transition from 700,000 annual sales of desktop computers in the late 1970s to tens of millions of PCs per annum barely a decade later is another; use of the Internet by billions of people would make anyone's short list of the most significant technological developments in recorded human history. In each instance, they would look for a combination of quantities of adoption and effects on human conduct.[11]

Often, a combination of discontinuities between one period and another signals to a historian that perhaps a new one exists with a new theme. For example, historians exploring the 1960s found such evidence. Popular music in the 1960s was significantly different than that of the 1950s, although the earlier genres were still played. Art underwent changes too, as Salvador Dali, Pablo Picasso, and others gave way to painters offering images of Campbell Soup cans. Long skirts were out; minis were in. Anti-Vietnam War protests and the Civil Rights movement displaced more conservative, comparatively calmer politics. University student populations exploded; economies improved dramatically; so did health, while birth control pills changed sexual behaviors almost overnight.[12] A student of modern societies, Michael Buckland, clearly framed the basic issue to address: "Any claim that our 'information society' is special or remarkable implies a contrast with some other 'non-information society' that is different in some noteworthy way."[13] But did older societies not also share information, talk, and distribute documents? These are questions historians work through when discussing the broad topic of information societies.

Normal practice calls for historians to describe criteria (some use the word characteristics) by which to judge features of an example to lead to a definition of a time. Without such criteria, anything goes. We could call our time the Age of the Potato because it is one of the most widely consumed foods in the Western world; or the Age of Rice, although that runs into an Asian-centered bias. But nobody in their right mind would accept the definition that we live in the Age of the Potato, just as perhaps a human many thousands of years ago would probably have scoffed at the notion that they were living in the Iron Age as they sat on a rock used as a seat by their fire where they now improved their digestion and health by cooking their meat.[14] For us to accept the idea that we live in an Information Age, which is sorely tempting to do, we would need some compelling criteria backed up by substantial evidence to justify that step. We need, too, as Michael Buckland argues, to perhaps acknowledge that "what is meant by an *information society* is that the way we live has become increasingly characterized by the use of documents in many forms."[15] He is right, of course, but as evidenced by our presentation, it is more than that, for it must include the shaping of information into forms useful to all manner of people and institutions. That is why becoming more tactical and methodical in discussing information and society is so essential.

To begin that conversation, I propose sixteen criteria with which to make a judgment about the Information Age issue that can be informed by events. We can just as easily consider these concerning features and behaviors within a society. These similarly parallel lists prepared by historians to judge the beginning and end of an era.[16] It is against these that one can build the case for or against the proposition that we live in an Information Age. The exercise is important because ultimately we face the blunt question, "So What?" or, even more rudely, "Who Cares?" We care because if our understanding of information and its role is better understood, then, as economists like to point out, we can leverage that to improve a society's economic productivity and quality of life.[17] Criteria are, as scientists argue, not as precise as they want—there is no set of overarching mathematical calculations involved—but a combination of qualitative and quantitative evidence that can be brought to bear on the subject. But a list for an era is a good place to start, as we move from fuzzy ideas about an Information Age to something more tangible, more specific, and more useful. To help keep the long list manageable, I grouped them into several larger categories: Institutional Responses, How much is created and used, and Economic and Social Impact. I do not suggest that some are more important than others, nor do I claim that this is the best possible list, but it is a start. That this list had to be so long is a declaration that naming an era is fraught with complexity and diversity of possibilities that must be accounted for if it is to endure as a reasonable name.

Institutional Responses

1. *Emergence of new professions appeared dependent on new bodies of information.* These professions included professional scientists, accountants, corporate managers, statisticians, specialties in medicine, over a dozen subspecialties in computer science, software programmers, computer repair personnel, airplane pilots and astronauts, and management consultants, among others.
2. *Expanded dependence on new bodies of information by preexisting professions.* While metal workers learned to use laser-cutting tools, blue-collar professions required increasing amounts of classroom training, such as for auto mechanics, electricians, and plumbers, and constant recertification for doctors, nurses, accountants, and teachers. Their reference books and manuals constantly changed and eventually went online. Computer diagnostics represented a new intrusion of automation. Sales of information media are measurable, such as the number of books and magazines sold, subscriptions to journals, PCs and smartphones, and fees for volume of data usage on smartphones.
3. *Existence of national laws and regulations to protect copyrights, privacy, and access to information.* As the creation of content of all manner increases, legal protections follow, such as copyright laws for phonograph records and CDs, movies and later television programming, and today software. This occurs when new industries and firms producing these lobby legislatures for such protections and other sectors of society that use these to block abuses (e.g., price gouging, exposing children to inappropriate content).
4. *Creation and expansion of information depots, such as university and public libraries, government, and nonprofit databases.* Some university libraries expanded from collections of 10,000 volumes before the 1860s to over 10 million a century later (e.g., Harvard University); national libraries did too in the United States, England, and France, for example. Governments and universities have some of the largest collections of

databases in the world, while thousands of associations and corporations maintain large ones too.
5. *Public administration that supports accessibility to information, education, free expression of opinions, and generally practices transparency in explaining its work.* Democracies do more of this than authoritarian governments. Political scientists and experts on media have documented changes in this behavior for the past two centuries, and historians increasingly for earlier times.

How Much Is Created and Used

1. *Formation of new bodies of information in addition to already existing ones.* Beginning in the mid-nineteenth century, new collections of facts began emerging in medicine, physics, biology, chemistry, mathematics, and statistics. These reinforced new lines of data creation and professions, stimulating new fields of academic study and professions. The number and variety of books, specialized journals, and magazines serve as further indicators of change.
2. *Reliances on new technologies are dramatic and measurable.* Inexpensive printing and telegraphy in the mid-1800s, later telephony, radio, TV, computers, PCs, and the Internet are obvious examples, but not the only ones. In particular, important advances in information technologies occur. TV and radio are almost universal; smartphones are used by over 50 percent of adults; satellites for transmitting radio and digital data become routine and used in commercial settings.
3. *Information content of work, credentialing, and licensing is extensive.* Studies of the evolution of specific professions are documenting this trend, including for jobs one does not think to consider, such as waitressing and bartending, carpentry, farming, and lawn care. Government reports for over a century have documented educational requirements, recording the increasing amount required across nearly 600 professions, including curricula for students.

Economic and Social Impact

1. *Formation of new businesses as a result of existence of new bodies of information.* Melvil Dewey's nineteenth-century card catalog business was a start, followed by vendors of file cabinets, 3 x 5 cards, adding and calculating machines, computer companies, and later Internet-born firms such as eBay, Google, Facebook, and Amazon. Traditional companies such as the *New York Times* established new lines of business, such as online editions and information retrieval services.

2. *Creation of new industries whose products and services are largely information.* These included the "office appliance" industry (firms selling cash registers, typewriters, and adding machines), later computers, PCs, iPads, and tablets; telecommunications (telephones, radio, TV, and smartphones); accounting (auditors, tax preparation); and real estate (home and commercial). Well over 200 new industries formed between the mid-1800s and the late 1900s just in the United States, similarly in Europe. These can be tracked by decade and size.
3. *Economic activities originating from information goods and services equals or that exceeds 20 percent of the GDP of a nation.* The pioneering work of economist Fritz Machlup in the 1950–1970s launched research on this aspect of information's role in modern economies. He included libraries and librarians, schools and teachers, universities and professors, and a vast array of government knowledge workers. Demographers and economists counted the new professions and the volume of revenue (or expense) they generated.
4. *Literacy levels exceed 75 percent of the population over the age of 12.* This is one of the best-documented developments in the industrialized world. Such levels were reached in many Western nations by the early 1900s. Today, literacy rates often exceed 90 percent. The number of students at each level (grades [American term] or years [British]) of education is well accounted for in almost every country on earth since the mid-1800s.
5. *Sales of information products exceed 10 percent of a nation's GDP.* This represents a shift from traditional services (e.g., waitressing, house cleaning, accounting, legal) and manufacturing to information providers, such as industry consultants, stock brokers, researchers at think tanks and industry associations, and even academic professional societies selling conferences, training, journals, and publications. This is about the commercialization of "knowledge work."
6. *Internet usage is above 10 percent.* Worldwide, by 2005, users exceeded 16 percent of the population; 48 percent in 2017, and 60 percent in 2021. In the developed world in 2005, extant records suggest over 50 percent had access to the Internet, over 80 percent in 2017, and in excess of 90 percent in 2021. The developed world achieved that 50 percent base target in 1997, the developing world one decade later, and the world as a whole in 2002. Everyone doubled those percentages in less than half a decade.[18] Clearly information exchange and content had changed from prior times.
7. *Popular culture includes texting, writing letters, blogging, and other forms of increased communication.* With the arrival of e-mail, then other forms of Internet-based communications, writing increased as

people wrote blogs, sent messages, and photographs regarding their thoughts and activities. Facebook is emblematic of that trend. In 2008, it had 150 million *active* users, 500 million in 2010, two years later one billion, and at the end of 2019, over two billion.[19]

8. *Societal values support scientific knowledge and practices, beliefs in the power of facts and truth, and openness to new ideas and data.* Survey data and academic studies of these issues document the growth (and decline) in these beliefs and values. During the 2010s, global debates over climate change made evident to what extent various societies valued these. Sometimes the values are perverse, such as in the dissemination of misinformation and lies over the Internet.

I am not sure that this long list is necessarily the right or complete, cataloging of characteristics of an Information Age, but possibly good enough at the moment for an information society or for a society filled with information. One could also move criteria about from one category to another. But no matter, it is a start because these characteristics allow one to catalog what and how much information appears, who creates and uses it, what institutions are involved, and the impact on an economy and the values of a society. To avoid the trap of calling ours the Age of the Potato, we need a broader set of criteria before naming an era. This is how historians do that properly.

ARGUMENTS FOR WHY WE LIVE IN AN INFORMATION AGE

Begin with economics. Two distinguished economists put the matter this way: "most of the increases in standards of living are," they bluntly announced,

> a result of increases in productivity—learning how to do things better. And if it is true that productivity is the result of learning and that productivity increases (learning) are endogenous, then a focal point of policy ought to be increasing learning within the economy; that is, increasing the ability and the incentives to learn, and learning how to learn, and then closing the knowledge gaps that separate the most productive firms in the economy from the rest. Therefore, *creating a learning society should be one of the major objectives of economic policy.*[20]

One of the two authors, Joseph E. Stiglitz, a Nobel laureate in economics, has long argued the case for why information increasingly became crucial to modern society, as did another of his colleagues who also is a Nobel laureate, Robert Solow.[21] The combination of a nation's ability to learn and to use its collections of information (they call it "endowment") is central to the prosperity of an economy.²²

Learning in this context is not just about students in school; more importantly, it is about research and development that converts into jobs and product and service innovations. Such activities occurred in all the professions and disciplines encountered in *All the Facts* and in *Birth of Modern Facts*. These economists move us from simply accumulating new and varied forms of relative information—they accept that—to taking the next step of using this endowment. They want further protections for intellectual properties, of course, education, and other incentives to further use of information. Prior experience indicates information's economic value had reached a point much earlier (in fact) where it was so extensive that state and national governments needed to continue maintaining an environment (social, legal, economic) that could sustain such information-based activities. Building on ideas originated by Joseph Schumpeter and others, the key to modern economic success is innovation, and that only comes from new knowledge, from inventing computer chips, for example, to knowing how to profitably manufacture them, and then learning how to use them in consumer and industrial products.

The second reason one might claim for why we live in an Information Age is that our dependence on information with which to do our work and engage in social and other personal activities has increased steadily in measurable ways since at least the eighteenth century in the Western world and in various parts of Asia, notably Japan since it began embracing various Western practices after 1868. By *measurable,* one can point to years of schooling, rates of literacy, numbers of books and newspapers sold, copyrights and patents issued by year and country, numbers of professional associations and each had, number and types of associations, specialized publications, and laws passed that supported intellectual property, freedom of speech, and the circulation of information. The percentage of information activities in each of these categories has increased continuously over time; the quantities and extent of their uses are measurable, too.²³ Arrayed against our list of sixteen characteristics, these arguments and quantitative evidence touch all of them.

Third, there is the qualitative historical record to draw upon. In one profession after another, discipline after discipline, and in historical and economic studies, there are testimonies of people complaining about information overloads, or reporting on efforts to organize and make information available in ever-larger amounts. Recall librarians, scientists, and business leaders creating, complaining, and controlling ever-growing amounts of information, data, and insights that kept transforming, too. Samples of that reportage peppered many books about modern times, including the knowledge worker literature, which increased the closer one came to the present.²⁴ Witnesses discussing the increased prevalence of information and how it was affecting the way one lived their lives were not limited to the United States, Great Britain, or France. The circumstances they described had become global certainly by the 1970s, while historians might push that chronology back in time.[25]

These arguments are reinforced by the approval of such points of view by historians, sociologists, members of the media industry, and others in various disciplines. Their evidence is extensive.[26] Here is how one student of the subject explained it: "Our political and economic, and social relationships, even our private ones, are more and more structured by what we call 'information,' to such an extent that we call the times in which we live 'The Information Age'."[27] Another reported that this belief made "the phenomenon of information" a "truly transforming and restructure catalyst."[28] The American sociologist, Daniel Bell (1919–2011), saw the post-industrial society as an information world;[29] the Spanish/Catalan sociologist and communications scholar, Manuel Castells, as transformed by technology and economics;[30] and American media critic and sociologist, Herbert Schiller (1919–2000), that this transformation dated to the start of the Second Industrial Revolution.[31] Since a field of study apparently does not become a recognized discipline unless it has a journal, the pro-information community acquired its own in 1981, *The Information Society*.[32] Barely a half decade later, one scholar crowed that the world had announced we had transformed from "an industrial society" into "an information society."[33]

That so many participants in the "Information Age" and their advocates are committed to this paradigmatic view of modern society makes it difficult to articulate rigorously what an information society even looks like, let alone do something similar to explain what an information age should be. One could argue that this problem exists as evidence of how fully engaged modern societies are in the belief that we live in an Information Age. But that can be discomforting for those studying the topic. As one commentator, Ronald E. Day, explains: "It is troubling because of its seeming naturalness and common sensibility and because of the ease of its predictions for an informing age of the present and the future."[34] Further, "its claims are far too simplistic and reductionistic of the complexities of sense, knowledge, and agency," leaving too many unfilled holes.[35] What happened can be explained, says this same observer, by the fact that "a tradition of values for information has been established and has been, rather uncritically and ahistorically, promulgated as a 'good' not only for Western culture but, more troubling, for, and as, 'the global'."[36] That is largely why arguments challenging the notion that we live in some special Information Age can be posited.

Then there are impressions one can add. Look around your apartment or home, and you will see digital information devices all around: digital TVs, iPads, smartphones, laptops, microwave ovens, programmable clocks and kitchen appliances, not to mention doorbells and Alexa, our newest digital fountain of information. Think about how much information one uses at work and play today—hours per day—then recall the situation in your home growing up. Clearly, you have more today than did your parents. If you are blessed to know your grandparents, think about what they had in their homes and at work—probably largely books and magazines, one or two radios, and a television set, and at the office, a typewriter, bookcase, and file cabinets. Think about how much organized information your parents had to process to do their work, and perhaps even your grandparents, and you will probably come to the conclusion that each succeeding generation was more dependent on organized information than the prior one. Research for *All the Facts* would prove you right.

ARGUMENTS FOR WHY WE DO *NOT* LIVE IN AN INFORMATION AGE

How much information do we need to be using across how much of society to conclude confidently that we live in an Information Age? Does having over 65 percent of the public using smartphones mean that we are far enough along to declare that we live in an information society? Has information transformed enough that the fundamental work of specific professions changed sufficiently to be so different from some pre-information society era? The last question is interesting because, for example, modern automotive manufacturers are extensive users of automation, numeric-control tools, robotics, modeling, and software managed supply chains, but these same companies do what they did a century ago: they hire and train people, make and sell cars. So, would we say automotive companies represent work in an information society? Yes, or would one have to see that they no longer made cars, but could still be in the transportation business, just making car-sized airplanes or rocket ships that one used to get around, like the American TV characters in *The Jetsons* in the 1960s?[37] So, the first argument against living in an information age is that we lack specific benchmarks for making such a decision.

An equally discomforting problem is that when we label this the Information Age, we abandon the opportunity to name it differently. There are other compelling options for labeling the era we live in. For example, Harvard University political scientist, Samuel P. Huntington (1927–2008), argued in the 1990s and early 2000s that the world was increasingly organizing itself into mega-civilizations characterized by shared attributes, notably religious ones. Islamic states were clustering together, Christian states were doing the same, and similar processes were underway in Asia. Cold War political alliances were gone (now possibly rebirthing in new forms), and new forms of tribalism were forming (e.g., democracies vs. dictatorships). He argued that politics and public policies were becoming increasingly culture centric.[38] The Middle Eastern wars that erupted after Huntington's predictions were between Christian and Islamic societies. Events in the 2010s and early 2020s affirmed his forecast in the United States of growing tribalism, too, perhaps turbocharged during the administration of President Donald J. Trump. Leaving aside the merits of Huntington's worldview, when he organized his thoughts, he paid scant attention to the role of information, even its technological superiority. Huntington predicted that less developed economies would catch up technologically. That development would bring everyone back to a balance of power and equivalent technical sophistication that had existed prior to 1500. Look at China today, and one sees that he was right.[39] Given the global problems of Islamic terrorism, 9/11, wars in the Middle East, and ethnic prejudice experienced by Chinese and Indian Muslims, we could just as easily label the same period as perhaps the Age of Religious Wars. What will historians in the late 2100s think, looking back? Was Huntington prescient or just stating the obvious, or possibly not? We are again back to the Age of the Potato issue.

A number of scholars view the issue less about whether we are or are not in such an information-centered era and more about debating the extent to which we are engaged with information and its consequences. They maintain that information has been in use for much longer than since the arrival of computers or the Internet. Herbert Schiller, and independently I, both concluded that many modern practices and dependencies on information stretched back at least to the nineteenth century, to the Victorian Era (English view) or the Gilded Age (American view).[40]

Jürgen Habermas would have us go back to the eighteenth century, as also another historian, Daniel Headrick.[41] Even further toward the "no" side of the ledger, British sociologist Anthony Giddens, has argued that all communities have been information societies from their origins.[42] British historian Edward Higgs, demonstrated that English public administration used information in ways similar to what is done by governments today, already in evidence in the early sixteenth century.[43] Others documented similar behaviors before the arrival of the Second Industrial Revolution.[44] Many of the features of modern society were partially implemented before World War I and that what occurred in the twentieth century was further deployment and increased transformation in information itself—the essential feature of much of my own research.

When does a community or nation flip the switch and become an information society? It is a question of degree and, more fundamentally one of specificity, of meeting some criteria, perhaps a more precisely defined collection of the sixteen proposed in this chapter. Historian Toni Weller has looked at this same issue, and I think she concluded correctly that "there is no obvious start or origin to the information society and, indeed, there is much work that suggests it is not a new phenomenon at all."[45] That reality makes it possible for historians to proceed the way they do best: by linking past developments to current realities. Daniel Headrick, an experienced student of the role of technologies and European history, agrees: "the Information Age has no beginning, for it is as old as humankind."[46] Given that historians have been documenting human history in considerable detail back several thousand years, Weller and Headrick seem to have the winning argument.

Before one rushes to any conclusions, however, one should also ask how the non-experts—society at large—thought about the matter. Is the issue one of either living in an information age or not? There is much work for historians to do to answer that question, but one can make a small contribution toward that research by briefly looking at two public discussions held when experts on information were just beginning to think about the role of information, data, and knowledge—they had yet to settle on the right language—in shaping modern society. To do that, we return to the 1960s and 1970s.

THE SHOCK OF FUTURE SHOCK AND THE FRENCH DEBATE ABOUT INFORMATION SOCIETIES

Scholars across multiple disciplines began to link rising quantities of information to changes in modern society in the 1950s. They noticed that increased quantities of information were straining people's ability to absorb it fast enough and to effectively take it in when making decisions.[47] While some controversy developed over who was the first to use the term "information overload,"[48] it was a self-trained futurist and journalist, with a B.A. degree in English from New York University (1950), Alvin Toffler (1928–2016), who brought it to the attention of the world. He did this first in an article published in 1965, but more spectacularly in a book in 1970, *Future Shock*.[49] Information overload was a concept he exposed to wide audiences in his "best-selling" book as part of a larger problem, which he referred to as "future shock." Toffler argued that this shock was the result of people, indeed entire societies, functioning in a psychological state where "too much change in too short a period of time" was occurring. The world had gone through two fundamental changes: first agricultural, then industrial, and now was in the midst of a post-industrial one, which was the subject of this and all his subsequent books.[50] As societies transformed from industrial forms to new ones relying on computers, more mobile work, and experiencing accelerating change, people felt disconnected and stressed, leading to many social problems such as the use of drugs, failed marriages, lack of happiness, and so forth. It is within that analysis of society's problems that he embedded the notion of information overload.[51]

Future Shock delivered an uncomfortable but rational explanation of changes then underway in society in well written if bombastic prose. It proved compelling. The book sold over six million copies by the time he died in 2016 and has been translated into over a dozen languages. It influenced the thinking of consultants, academics, and senior public officials around the world because they saw that science, capital, and communications were converging, causing many changes. His obituary in the *New York Times* noted that "he was among the first authors to recognize that knowledge, not labor and raw materials, would become the most important economic resource of advanced societies."[52]

The public on both sides of the Atlantic, in Japan and among Latin Americans, including university students, senior public officials, and corporate leaders, accepted his analysis, seeing these as compatible with their own less organized observations about what was happening in their lives. On North American campuses, seminars and lectures devoted attention to his ideas. His subsequent books continued the same themes. These were discussed in China as it was emerging from its Maoist economic practices and in Japan as it was transforming its social constructs. In an interview with the *New York Times* in 2006, he advised readers to focus more on "the rate of change," because that "has implications quite apart from, and sometimes more important than the directions of change."[53] People debated what changes were coming, with his book serving as their first significant introduction to ideas of the Information Age already animating sociologists and economists. It was also all quite unnerving because "the concept of future shock and the theory of the adaptive" was "not the final word, but as a first approximation of the new realities, filled with danger and promise, created by the accelerative thrust."[54] The public had not encountered such a message before. It really was shocking.

To deal with this new reality, Toffler led with informational solutions. Future fitness called for literacy as a form of resilience, but it needed to be grounded in values suitable to the new age. To be adaptive, people had to cultivate curiosity, especially to support personal growth and economic development, points made also, for example, by Joseph Stiglitz and Robert Solow. Success in this new world would be driven by a healthy dose of new information and a willingness to be open to new ideas and facts. Information and ideas about information societies and eras were now permanently public issues of discussion across the industrialized world. In the process, Toffler became a household name. *Future Shock* remains in print, and he is still discussed in academic circles a half century after first publishing his book.

In France, society and jobs were also changing. Businesses and governments were increasingly using computers for the same reasons as in the United States and all over Europe.[55] The same academic disciplines that were changing the shape of information in the United States were simultaneously doing the same in France and in most parts of Western Europe. Blue-collar workers became nervous over the possibilities of automation limiting their work prospects. A special concern within French government and private circles was the growing dominance of American computer companies within the local economy, notably IBM.[56] Fretting about the nation's inability to develop a local computer industry weighed on the administrations of Charles de Gaulle (1890–1970), who served as president of France (1959–1969), and Valéry Giscard d'Estaing (1926–2020), president from 1974 to 1981. During both their terms, reliance on information in all walks of life increased, levels of education did too, and the role of data in daily work practices raised the visibility of changing informational circumstances in France.[57] President d'Estaing was keen on modernizing his economy based on such technologies as nuclear power and computing. By European standards, he was politically conservative but administratively innovative and open to new ideas and social norms.[58]

In 1976, he commissioned a study on how computers were affecting society, and specifically how the American industry was doing so in France. On its surface, such government studies are benign exercises anywhere in the world, but useful in informing senior public officials about issues of the day. In this particular instance, he asked two accomplished individuals with both public and private experiences to lead the study, which suggests that he wanted this report to engage the nation and its political factions in a broader debate about technology than merely for an elite group of Parisian thought leaders. The first of the two authors, Simon Nora (1921–2006) had trained in law and, after World War II, rose quickly through the ranks of the national government, largely in financial and economic positions. He was also a speechwriter for senior French officials. Just as important for our story, he served as the general manager of a leading French publisher, Hachette, in the early 1970s, which launched a weekly publication, *Le Point.* So, he brought to d'Estaing's project public sector experience, the ability to write, and media/publishing savviness. His co-author, Alain Minc (b. 1949), had a similar background, combining public and private careers. He also had business consulting experience, served as the Chief Financial Officer of the Compagnie de Saint-Gobain, a manufacturing firm, and was considered a rising star in French business circles. In the years following his collaboration with Nora, he wrote over two dozen books.[59] So these were not two faceless government staff members commissioned to write a dull report destined to gather dust.

When published in 1978, their report, *L'informatisation de la société*, nearly instantly became a "best seller" in France.[60] Published under the imprimatur of the government's publishing arm with a bright red cover in a multi-volume edition, the most popular piece of the report was volume one, which today one would call the executive summary. It was soon republished in other languages, each with an introductory essay by a leading expert. The English-language edition, for example, published by MIT Press in the United States, had as its guest essayist the prominent sociologist Daniel Bell.[61] These two French authors knew clearly how to disseminate their message.

The heart of their study was a consideration of the threat IBM and other foreign nationals posed to French society. They laid out possible visions of the future of society, the part of the report that triggered considerable dialogue in the press, university classes, and television.[62] The foreign computer invasion of France would appeal to conservative nationalists, who were always wary of external cultural influences on the nation, while progressives were drawn to speculations about the future of society at a time when it was undergoing much change due to such issues as abortion, divorce, and the role of France in the Cold War. The topic represented a continuum in business circles encouraging further computerization—hence use of information—in their operations.

The report called for a national strategy for how society should use computers, of course, but in the process of laying out that case, the authors introduced a new word into the French language that galvanized the discussion as much in France (and later Europe) as "Future Shock" had in the United States: *télématique.* Loosely translated, it referred to the combined uses of telecommunications and computing and most famously led to the development of a terminal that served as a telephone as well, for use in homes and offices. Called the *Minitel,* it became available starting in 1981, and proved popular in the years before the Internet.[63] The authors argued that ICTs (information and computing technologies) would reshape society, and it was that contention that led so many in France to pause and consider how information (and computing, too) was doing so. Readers debated the report's ideas and asked questions concerning issues such as privacy of their information, the role of automation in disrupting lives, of improving them too, of disparities in the French educational system hence career opportunities, and attitudes toward sex, religion, and nationalism. Their discussion extended for years after the publication of the report.

Meanwhile, the government launched several initiatives to encourage the expansion of a local computer industry and to block IBM, a story that need not distract us here.[64] The French saw these initiatives as a way of protecting the sovereignty of their nation—always a concern—which mixed with debates over the changing nature of society. The primary effect of the report was similar to Toffler's: it cemented in the public's mind that they were beginning to live in a new era, an information age. The daily reminder came in the 1980s with the presence of the *Minitel* terminal in their homes and offices.

IMPLICATIONS

When Toffler, Nora, Minc, and so many others wrestled with the issue of the obviously increasing use of information in the 1950s-1980s, they could be excused for entwining into their deliberations the obvious intrusion of computers into the buildings of government, business, and universities, and later into smaller organizations and into their homes. France's telephone terminal, *Minitel*, was only a temporary, if dramatic, material symbol of what was becoming apparent: that electrified information (both analogue and digital) was intruding into professional and private lives much like a storm approaching or more silently, like a bad cold spreading through one's household. These circumstances were their realities, ours too. But, what would the actors who walked through the pages of *All the Facts* and *Birth of Modern Facts* have added to the conversation?

Historians would argue that each, in their time relative to what came before them, was shaped by their current social, political, economic, and intellectual transformations of their day. All had also lived in an age abounding with almost out-of-control quantities of information crushing them. As rates of literacy began to rise, beginning widely in the sixteenth century in Europe, they probably would have joined colleagues in the twentieth century to argue that they, too, lived in an information age, if we had to pick a label. But Germans in the 1600s would have protested because the Thirty Years War (1618–1648) was comparable in importance and brutality to the two twentieth century world wars (1914–1945) that almost share the exact dates three hundred years apart. Yet, one would be pressed to find any serious student of the period 1914–1945 who thinks those years should be labeled an Information Age, for the same reasons their ancestors in the 1600s would not. People living in the eighteenth century experienced constant wars, too, largely among the French, British, and Spanish, but everywhere else too, if in differing amounts. The First Industrial Revolution flowered, bringing people to factory jobs off the farms, causing cities to expand rapidly, diseases to infect farm animals and urban dwellers, all while literacy rates were increasing and more people began receiving a modicum of school-based education. By the end of the 1700s, merchants who imported or sold goods to other parts of their country or shipped to other nations could not avoid understanding in written form laws governing trade and accounting, hence mathematics. If they sailed their own ships, that meant, too, learning some geometry.

So, we can reach the conclusion that people in many periods in the past were enveloped with more information than their ancestors; that they experienced contending dramatic events that somehow always outranked in importance considerations about information ages; that they knew written-down facts were important and influential too, but not commanding. For information domination would require a force at work greater than, say, war or the move from agricultural to industrial economies. That could come, however, if AI marries up with robotics and comes to dominate activity and energy use on Earth. But until traditionally more important human actions are overcome by AI and robotics, or by some other biological forms yet to dictate events on Earth, Information Age identities will probably play a secondary role.

Does that mean thinking of the elements that go into promoting the candidacy of Information Age for our time is a useless exercise? The effort is worth it for the same reasons that, in earlier times, knowledge workers organized information and developed new ways to use it, even eager to create new data to reinforce such aspirations. The arguments for and against the new name and our list of sixteen characteristics are less about defending a new one than they are about intellectual tools to help understand how to use existing information. Ultimately, this is what earlier knowledge workers wanted to do as well. Entwining computers into the effort, as did Toffler and the French, is exactly what is needed, because computing is today's tool that has been used sufficiently to create, manage, and help humans go about their lives.

CONCLUSIONS

So, let discussions about the Ages that we live in continue; these are useful and indeed important. Expectant parents spend months and hours upon hours debating what names to give their soon-to-arrive baby. One would be hard pressed to find exceptions to this practice. They worry about what signals regarding behavior a name will send to their unborn child. The father worries about whether the baby will get teased on a school playground a decade later. The mother worries about whom in the family she is blessing or irritating with a name. Naming is rife with social pitfalls, too. A Catholic baby in the 1700s not named after a saint guaranteed the parents would receive a scolding from the local priest as he denied baptism to the child, ensuring that the little one would go straight to purgatory. When IBM named a PC the "PCjr" and it became a failed product, one was guaranteed that never would any computer company dare use the term "jr" in the name of a machine. In the end, parents normally pick names that work within the context of their times, even though their children turn out "good" or "bad" based more on how they were raised or the economic or social environments into which they were born. The lesson for those concerned about information ages is to debate the issues, but not to take them too seriously, because bigger circumstances are at play.

Nowhere does this seem more so than with two issues: the roles of Big Data and fake facts. These two developments, while they have long histories—dating to the first instance at least of Paul Otlet (1868–1944) and his millions of tabulating cards to document publications before World War II and the arrival of computers, and the second to the lies of politics back hundreds of years—are obvious and important features of contemporary society. Both became important because of the availability of the Internet and massive computing and the participation of virtually anyone who had access to the Internet. These two circumstances can almost be said to be new enough in their currently virulent forms, despite their ancestry. Both, thus, draw our attention, each deserving a chapter. We turn first to Big Data because its presence is more ubiquitous than the more obvious and noisy flow of fake facts through modern society.

Chapter 4

The Emergence of Big Data

The world contains an unimaginably vast amount of digital information which is getting ever vaster more rapidly . . . The effect is being felt everywhere, from business to science, from government to the arts. Scientists and computer engineers have coined a new term for the phenomenon: "big data".

—Kenneth Cukier, 2010[1]

Is data the new oil?

—Perry Rotella, 2012[2]

In the first half of 2020, many millions of people around the world learned a new phrase: "flattening the curve." Doctors, scientists, and public officials at all levels of government used it. It referred to inhibiting increases in COVID-19 cases beyond the capabilities of hospitals to treat patients. Media explanations of this as a strategy to slow the spread of the epidemic were accompanied by graphics that looked like two mountains, one high and the second low below a certain line, with that second line representing the capacity of a hospital (or nation) to properly handle patient loads. To make sense of the "flattening the curve" strategy, the public and officials were introduced to statistical methods and models used to predict how many outbreaks there would be and by when. This was a conversation about infection rates rising and falling, predictions about the lack of sufficient ventilators and masks, and the number of anticipated deaths. It would be difficult to exaggerate how many people saw the media coverage of these models, their harrowing statistics, or explanations of public health strategies to slow the pandemic's spread. What hundreds of millions of people were exposed to was a case of

Big Data. The public paid attention.[3] The dissemination of such facts proved useful in controlling the spread of the virus, presenting a positive example of massive quantities of information and predictive analytics merged with proven medical strategies.

But then, there was Facebook, also using similar software tools and models, collecting and analyzing vast bodies of data but with controversial outcomes. It accumulates, analyzes, and sells information about literally billions of people. Google and Amazon do, too. I discuss these firms later, but here it is important to understand that they continuously track people's activities and what they record and write, making surveillance the "killer app" of Big Data. Facebook has a history of poorly protecting data about individuals, resulting in data breaches. European and American government agencies have explored these breaches, concerned about the security and privacy of individuals' information. They have fined the company. In 2020–2021, the European Union, the U.S. Federal Trade Commission, U.S. Congressional committees, and the U.S. Department of Justice studied Facebook's behavior with an eye cast on whether it had violated laws and regulations. The most notorious instance was the Cambridge Analytica data breach scandal. In 2018, then Facebook, now called Meta, acknowledged that an "app" developed by Global Science Research harvested personal data on up to 87 million Facebook users without their consent. The news created a furor on both sides of the Atlantic, raised concerns about privacy issues, and triggered various investigations.[4] Facebook reported that the data included an individual's public profile, page likes, their birthdates, and locations.[5] Additional data reportedly was also accessed, such as profiles of one's News Feed, timeline, and messages (if the individuals accented). Cambridge Analytica collected sufficient data with which to construct psychological profiles of individuals to optimize political advertising aimed at them.[6] This was a negative example of Big Data.

During the pandemic, a third case of Big Data at work captured momentarily much public attention: the use of Internet-connected thermometers. Kinsa Health manufactures and sells "smart" thermometers to consumers that track one's temperature, collecting this data on over two million consumers. Take your temperature and the firm knows when you did that, how much fever you had, when, and where. The company uses that data to predict the spread of flu, for example, by spotting rising temperatures. In 2020, it began to do the same for COVID-19, predicting where it might come next because an early symptom of infection is fever.[7] For a number of years, it accurately predicted flu outbreaks by two weeks in advance and where these were occurring. It proved able to do this even before public health officials could at the Centers for Disease Control and Prevention (CDC) in the United States or by local agencies. The general media now published its maps.[8] With years of data by date and location, if there was a spike when compared to what normally were rates of fever on a day or place, it could predict (anticipate) flu outbreaks. It was now signaling when possible outbreaks of COVID-19 (or some other illness) would come, say in South Florida, which proved accurate.[9] This data represented a new body of information resulting in novel ways to conduct disease surveillance. Such tools held out the potential for public health officials to swoop in earlier into a potential hot spot to test and treat an outbreak. This was another case of Big Data at work.

In each case, vast quantities of data had been collected with the use of software and sensors involving minimal human participation. These data were then analyzed, insights gained, and actions taken. Facebook and Google used it to sell information to advertisers about who might find their products attractive. Facebook made it possible to allow software developers to tap into those databases in potentially unethical ways that at a minimum seemed like sloppy data practices.[10] The thermometers opened up new possibilities that other large bodies of data offered, such as the massive accumulation of information about climate change (e.g., atmospheric pollution, ocean temperatures, shrinking of ice in the polar regions). Many trillions of pieces of data are involved, massively more than the computerized databases of the twentieth century. Big Data emerged as a new class of information, mostly associated with the twenty-first century, but roaming around our information ecosystems by the mid-twentieth century. As with all new developments involving computing, the hubris started glorifying its benefits for humanity. Experts were proclaiming that "The era of Big Data has begun."[11]

Hubris is both a motivating and pejorative idea that has dogged the trail of information technologies and digital data from the beginning. Hubris is good in calling attention to a new tool, body of information, or use. Extolling the virtues, say, of a Covid vaccine in sufficiently loud and positive voices may encourage some individuals to be inoculated, thus potentially saving lives or increasing one's understanding of the dynamics of Covid. But hubris can also overstate a case or hide negative considerations. Some observers point out that just staring at vast quantities of data does not necessarily reveal truth. Examining information created by Big Data does not necessarily mean that an obvious product of such an exercise—correlations—is enough to reveal a new truth. Such exercises do not result in such findings being "the magic of big data." That magic suggests, as one expert describes it, that "You don't really need to know or understand anything about what you're studying; you can simply place all of your faith in the emergent truth of digital information."

That can be considered a Big Data fallacy, of course, or to put it in less prosaic terms, as nonsense or intellectual risky behavior, and in more academic language, "an outcome of scientific reductionism." In other words, it can be the acceptance of a "belief that complex systems can be understood by dismantling them into their constituent pieces and studying each in isolation." Again, one does not need a PhD in the hard sciences to realize this can lead to a bad case of Nonsense.[12] This is where Ernest Hemingway's admonition to always have turned on one's "Crap Detector," is the prudent way to approach some Big Data and loud and noisy hubris. Our book tends to see the positive in the role and evolution of information, the idea of the glass being half full, but that optimistic inclination is not unlimited, for everything has its downsides that one should consider, and information and its roles are no exceptions.

As with all types of information, facts have a history, but now, too, a future that is already present. That is why this chapter introduces Big Data, a new class of information, as a new feature of modern life. I introduce features of Big Data, their emerging impact on society by how it is used, and its implications. It is not too early to point out that the constituencies discussed by historians of academic disciplines and business/government are the same developing and using Big Data and are collaborating across their areas of expertise in greater numbers in part because of the availability of such collections of information. This chapter begins with a definition and description of Big Data. Then a discussion follows about its uses and implications for society. The availability of Big Data is affecting how users think about the role of information (data), so that issue is addressed, based on what has been learned so far about the role of Big Data and prior collections of information.

DEFINING BIG DATA AND ITS EMERGING FEATURES

"Big Data" as a term has existed since the 1990s; however, as a recognizable one, it is realistically a twenty-first-century expression, even though more anonymously evident as a concept by the 1950s.[13] It was more than just very large files that exceeded the design and capacities of twentieth-century database software tools, although that is one of its distinguishing features. Traditional analytical and database management tools could not handle the volume. Big Data requires different methods for collecting information, such as a continuous inflow of, say, meteorological data, or stock market information at sub-second speed, or managing the volume, which can be orders of magnitude more than with prior data sets. New methods were required to share, move, transfer, analyze, visualize, query, delete, vacuum up disparate files, update, and manage privacy and security. By the second decade of the new century, definitions still remained tethered to size and technology: "Big Data is where parallel computing tools are needed to handle data," thus is a "change in the computer science used" from what had been applied before to relational databases.[14]

If all this sounds fuzzy, it is because consensus has yet to coalesce around what the term means, so various disciplines have served up definitions, descriptions of its scope and features, and guidelines on how it should be used.[15] Increasingly, definitions tend to describe it as a category of quantitatively oriented products, cognition oriented, looking at how people relate to data;[16] and process oriented, focusing on how data is collected, analyzed, and curated.[17]

Facebook (Meta) takes a product-centric view, collecting many hundreds of terabytes of data on people.[18] Twitter (now X), Google (also known as Alphabet), and Amazon similarly collect terabytes of data. Doug Laney, a consultant, in 2001 famously presented the central feature discussed often about this orientation.[19] He argued that Big Data evolves along three dimensions: volume, velocity, and variety. Others added value and veracity to his list.[20] About data volume, Laney pointed out that firms were collecting a great deal more, but less expensively than in the past. A clear consequence is that they "become reluctant to discard it." Data velocity was a result of increased e-commerce, which "increased point-of-interaction (POI) speed and, consequently, the pace data used to support interactions and generated by interactions." As its value increased as a competitive edge, access to and analysis of that data sped up. Regarding data variety, he saw it increasing as preexisting data sets in differing databases with various data structures and so forth meant Big Data had to be some mega-controller over multiple data sets. It had to provide "change/translation mechanisms" to control the "portfolio sprawl" evident at the turn of the century. But the variety was always going to be a feature of Big Data.[21] His last point can be overlooked if we are not attentive to the details because this represented a significant departure from prior notions of databases. In the twentieth century, databases tended to contain the same kind of information within each, not various collections. Big Data often had various collections that somehow had to be connected and analyzed together. For Laney, Big Data was a large data management topic, and he focused on what needed to be done to make these files useful.

At the risk of deviating from our discussion about how information changed via Big Data, at least a listing of new tasks begins to suggest how Big Data differed from non-Big Data. Regarding large *volumes*, one would normally need to organize these in both tiered storage and in hub and spoke organizations of information. Other activities would be needed as well: selective data retention, statistical sampling, elimination of redundancy, offloading "cold" data, and outsourcing. To deal with *velocity*, one would need to use operational data stores, data caches, point-to-point data routing, and balance data latency with decision cycles. These are techniques honed decades before but in need of tailoring. Finally, regarding how to manage *variety*, Big Data required resolving inconsistencies in the data, using XML-based "universal translation" tools, embracing application-aware enterprise application integration adapters, data access middleware, distributed query management tools, and applying well-understood metadata management software and methods.[22]

Regarding humans—the cognition issue—as one observer noted, the problem was that Big Data's "interactions are so complex and massive that human brains simply cannot comprehend them."[23] That circumstance called for different software mediation, statistical analysis, assistance from brain experts, and new forms of visualization to help us mere mortals.[24] Big Data also concerns the application of predictive analytics and how users behave as both pattern recognizers and predictors, which is what Facebook and Amazon need done to sell useful information about people to advertisers and vendors. Analyses of large collections of data are conducted to identify new trends, correlations, dangers, and problems, and so forth that smaller data sets would not reveal.[25] As Melanie Feinberg, an information sciences expert, explains, Big Data's attraction is its "allure for certainty," which the mix of computation and its ability to work with large volumes of data makes so attractive across all the hard sciences and many social sciences, too.[26]

So size and growth are central to the notion of Big Data. These issues are the results of inexpensive automated, often sensor-based data collection capabilities that do not involve humans. These include the use of cameras, RFID tags, wireless sensors, Internet of Things (IoT) devices, and manufacturing and process machines (even automobiles) themselves all sending in information. It is within this context that people read so much about how many exabytes of data are collected each day.[27] While all this is fascinating, once we reach a situation where one has a great deal of data, how many times greater after that does not fundamentally change what Big Data is about.

Veracity is about accuracy. Observers treat this as unique to Big Data because one is concerned about the accuracy of automated data collection and analysis.[28] However, this is an issue with any set of data in any period of history. The final *V* is routinely served up as a feature of Big Data is *value*, which is the utility gleaned from data. Not for a lack of *V* words, there is *variability*, too, which acknowledges that Big Data's value or other features evolve as the context shifts in which they are created and used.

One feature—perhaps an influence—of Big Data on computer scientists, practitioners, and students of information is that it created a heightened sense of differentiation between the terms *data* and *information*. Throughout my studies, I have used the terms rather loosely, occasionally interchangeably, but in the twenty-first century use of their content, these words increasingly reflect a looseness that is sloppy. With so many signals (to use Claude E. Shannon's notion) now constituting what is actually Big Data, we must think of information as the anthropomorphic extractions from those files in terms humans can understand. We now live in a world in which data outnumber information. For example, Facebook may have 5,000 to 6,000 pieces of data about an individual, but if that person were to examine these, they would be shown information, and that quantity would be expressed in a human language and be far less than 6,000 facts.

A bit of history reveals that Big Data is part of a longer evolution in the nature of information, a facet, however, that was substantively an issue of ever-bigger files. Several data points demonstrate this link to earlier computing and digital and analog file management. Computer scientists worried in the 1960s that data storage might exceed storage capacity in computer systems, so they urged that "storage requirements for all information be kept to a minimum."[29] To the mid-1970s, technological innovations had kept up with demand and actually increased it, so national governments, computer scientists, and IT vendors no longer blocked the growth in data; instead, they kept up with it and knew how to use it.[30] Yet, by the mid-1980s, their concern of, "Can Users Really Absorb Data at Today's Rates?" had not gone away.[31] Even computer scientist Peter J. Denning, always the optimist about computing's future, as late as 1990 worried that "The imperative to save all the bits forces us into an impossible situation: The rate and volume of information flow overwhelm our networks, storage devices and retrieval systems, as well as the human capacity for comprehension," which current technology could not handle.[32]

By the late 1990s, possible solutions to the problem were being proffered, and just as important, the term *big data* began to be used.[33] By the end of the decade, sensors and other sources were transmitting data over the Internet, forcing new procedures to be formulated for handling this influx of new types of data. Big Data had started to seep through the computer science community as one of the next technological challenges to be faced. In 2001, Laney published one of the first commentaries on how to deal with the problem.[34] Forecasts of ever-growing files continued to appear, but increasingly, too, commentary on how Big Data was changing the way organizations operated. For example, in 2008, several experts concluded that just as search engines transformed how one accessed information, other forms of big data computing could transform the activities of companies, scientific researchers, medical practitioners, and a nation's defense and intelligence operations. They added what must have seemed like hubris at the time, but not so a decade later: "Big-data computing is perhaps the biggest innovation in computing in the last decade. We have only begun to see its potential to collect, organize, and process data in all walks of life."[35]

If we step over the now usual commentary about how the volume of data was expanding rapidly, one can see that these scientists memorialized changes already emerging in the nature of Big Data information uses. They offered examples:

Wal-Mart recently contracted with Hewlett Packard to construct a data warehouse capable of storing 4 petabytes (4,000 trillion bytes) of data, representing every single purchase recorded by their point-of-sale terminals (around 267 million transactions per day) at their 6,000 stores worldwide. By applying machine learning to this data, they can detect patterns indicating the effectiveness of their pricing strategies and advertising campaigns, and better manage their inventory and supply chains.

Many scientific disciplines have become data-driven. For example, a modern telescope is really just a very large digital camera. The proposed Large Synoptic Survey Telescope (LSST) will scan the sky from a mountaintop in Chile, recording 30 trillion bytes of image data every day—a data volume equal to two entire Sloan Digital Sky Surveys daily! Astronomers will apply massive computing power to this data to probe the origins of our universe. The Large Hadron Collider (LHC), a particle accelerator that will revolutionize our understanding of the workings of the Universe, will generate 60 terabytes of data per day—15 petabytes (15 million gigabytes) annually. Similar eScience projects are proposed or underway in a wide variety of other disciplines, from biology to environmental science to oceanography. These projects generate such enormous data sets that automated analysis is required.[36]

Modern medicine collects vast amounts of information about patients through imaging technology (CAT scans, MRI), genetic analysis (DNA microarrays), and other forms of diagnostic equipment. By applying data mining to data sets for large numbers of patients, medical researchers are gaining fundamental insights into the genetic and environmental causes of diseases and creating more effective means of diagnosis.[37]

The collection of all documents on the World Wide Web (several hundred trillion bytes of text) is proving to be a corpus that can be mined and processed in many different ways. For example, language translation programs can be guided by statistical language models generated by analyzing billions of documents in the source and target languages, as well as multilingual documents such as the minutes of the United Nations. Specialized web crawlers scan for documents at different reading levels to aid English-language education for first graders to adults. Current generative AI is "learning" how to respond to queries and create new text, images, and videos, by surfing the web of vast collections of Big Data. Each of these uses of Big Data became a reality before the ink was barely dry on their report.[38]

But perhaps the world was beginning to settle into a workable perspective by 2012 when two computer scientists referred to Big Data as much a cultural and scholarly phenomenon as a technological one. They added the further twist of "Mythology: the widespread belief that large data sets offer a higher form of intelligence and knowledge that can generate insights that were previously impossible, with the aura of truth, objectivity, and accuracy."[39] The voice of experience with Big Data was beginning to be heard. That could only happen after people had moved from definitions of Big Data to working with it.

One other feature of Big Data needs recognition, because it is evident in multiple types of information: the difference between discovering and creating data (information). Information experts are increasingly describing today's information as facts created—the idea of *creation*, or invention of new facts. For example, when a city wants to know how many vehicles drive on a specific road and at what hours, it lays down a wire across the road that, when one drives over it, counts that event. Thus, the city has created a body of new information that did not exist before, latent in nature, for example, but here made as a consequence of humans developing roads and vehicles. With *discovery*, the purpose is to explain (understand) a preexisting natural, scientific, or other situation, such as when a physicist discovers what kind of atmosphere surrounds a planet, or when a medical researcher finally identifies how a particular gene causes cancer. The more a fact emerges from hard sciences, the more likely it is to be a discovery. The more that a fact is an observation or new creation, the more likely it came from the social sciences, such as our observations about the nature of information explained in the first chapter of this book. All disciplines and all subject areas tend to have a combination of discovered and created facts. Fake facts are an example of created information too, so relevant to our study of contemporary information. And, of course, using the word *creating* as a verb is the act of pulling together various facts to make sense of a situation, such as the amount of traffic on a road. Finally, much of information's change and growth in the nineteenth and twentieth centuries were about the discovery of physical realities about how the world operates, or did, and the role of humans. By the end of the twentieth century, however, humans were also creating new bodies of information at the same time.

So, data evolved from being facts floating around outside the context of some process of discovery to being the preferred source of much experimentation and observation today. It could be accurate (truthful) or not, or not even understandable and meaningful. It could also be used as part of the act of creation of new information, again either accurate or not, tied to some context or purpose, or not. In short, it portends possibly new changes in future information.[40]

IS THERE A CIVIL WAR UNDERWAY AMONG BIG DATA USERS?

In the world of data, information, and knowledge, there have always been epistemological and methodological debates. For the most part, these were quasi-scholarly differences of opinion about the nature of facts; discourses intellectual luminaries engaged in during the nineteenth century and continued to conduct until mathematicians, engineers, and radio ham operators started electrifying information with computers. Then these debates became blunter, less polite, and, of course, involved the creation of entire industries that in time came to consume more than 10 percent of the world's economy (GDP). The fundamental fault line that seemed like clashing tectonic plates was the divide between qualitative and quantitative data, with information, facts, and knowledge being claimed by one side or another. The debate never subsided, although peaceful gestures were made by the social scientists when historians, for example, partially embraced cliometrics. The hard scientists refused to give up a centimeter, while the humanists were almost as stubborn, too.[41] But the latter were outnumbered, as there simply were not enough literature and language graduates or philosophers to take on the turncoats in the social sciences. The real problem was that both qualitative and quantitative information had evolved for all the reasons trotted out in previous chapters.

The arrival of Big Data set off another round of debates about these tectonic plates while most users went about leveraging the new data sets in many disciplines. So what has Big Data contributed to the informational civil war so far? We must say, "so far," because despite the hubris around Big Data, it is an innovation that is just, if rapidly, gaining a permanent foothold in the larger world of information. This is occurring despite the rapid and spectacular successes of Facebook (Meta), Twitter (X), Google, eBay, Amazon, IBM, Microsoft, and Apple among the leading lights of this new world.

What happens if a humanist or social scientist gains access to large bodies of statistics or other "signals"? Does "playing with data" then "serve as a gateway drug" that pulls them away from qualitative information, hence the rich contextual insights and stories that so inform their work, and into a world of cold numeracy?[42] Historians of technology always caution each other to be careful of "technological determinism," the idea that technologies force events to occur in prescribed ways, detracting from qualitative influence on activities by people—related to the idea that regardless of circumstance or technologies, people can make decisions and take actions.[43] On the other hand, proponents of using digitized files argue that these are tools that enhance the core work of the qualitative historian.[44] I subscribe to that latter school of thought. Looking at a Big Data file while working with large multinational companies and government agencies did not convert me into a quantitative historian; it merely forced me to persuade statistical mavens to ask questions of the data on my behalf.

So far, most of the "data-diggers" live in the hard sciences along with economists, but are increasingly being joined by sociologists and political scientists, among others. Experts on working with Big Data increased in number after 2000 and are embedded in all manner of disciplines, organizations, and industries. This development was facilitated by the continuing drop in the cost of computing, increases in the power of computing, the now-existing collection of data mining techniques developed in the 1980s and 1990s, along with methods for predictive analytics and, of course, creating and manipulating Big Data files. Two observers of such developments summarized what was now happening: "a variety of service providers and start-up companies were born to sell digital packages to the new miners, and the gospel of data science filled their marketing materials, turning the flywheel of data exuberance and encouraging companies old and new to reconsider their data strategy and staffing."[45] Big Data did not cause any discipline to embrace vast files; rather, their earlier moves to mathematics and statistics and databases of the 1980s and 1990s did. But behavior changes. Just as I had to find statisticians at IBM, so too to leverage Big Data in other disciplines and institutions required multiple skills and collaboration across disciplines. A full range of new methods of research and analysis had arrived and been adopted.

As one observer of this unfolding process pointed out, "Scholars in all fields are taking advantage of the wealth of online information, tools, and services to ask new questions, create new kinds of scholarly products, and reach new audiences."[46] Many in academia and research positions in corporations and government see the availability of such massive new collections of data as useful in uncovering information that leads to innovations.[47] The specter of technological determinism still exists but can also be seen as a positive influence that has its advocates. A professor of information makes the case:

> The volume of scientific data being generated by highly instrumented research projects (linear accelerators, sensor networks, satellites, seismographs, etc.) is so great that it can only be captured and managed using information technology. Social scientists are analyzing ever-larger volumes of data from government statistics, online surveys, and behavioral models. Humanities scholars, similarly, are producing and analyzing larger bodies of texts, digital images and video, and models of historic sites. The amount of data produced far exceeds the capabilities of manual techniques for data management. This "data deluge" is pushing efforts to build an advanced information infrastructure for research and learning.[48]

It is a familiar repetition of events: create new information and people will want it, use it, and develop tools and methods for doing so. It is an old story, one encountered as far back as with the librarians and their cards in the eighteenth and nineteenth centuries. As in earlier times, the usefulness of the dichotomy of qualitative and quantitative comes into question when the two can be blended, even fused because of the diversity of tools and data now available. Questions of methodology, to quote a team of scholars looking at the role of Big Data, "that arise, therefore, derive from the stage at which subjectivity is introduced into the process: that is, the decisions made in terms of sampling, cleaning, and statistical analysis. Proponents of a "pure' quantitative approach herald the autonomy of the data from subjectivity." Also that "human intervention at each step undermines the notion of a purely objective Big Data science."[49] This is what I experienced, for example, by going to statisticians at IBM with a list of questions, guiding them on how they should "massage the data" to extract answers. In the process, they suggested other "interesting" lines of investigation made possible by the particular organization of the data in question.[50]

The methodological boundary is a familiar one: human subjectivity imposed on shaping the features and value of data, information, and insights. The reaction to how best to work at this frontier plays out in debates about qualitative vs. quantitative information.[51] Raw data in any quantity is normally not organized the way someone wants in order to use it, so it is reorganized to fit one's needs. The data miners "clean" or "condition" it, to use the jargon of Big Data users. What elements of a data file should one look at and which to ignore is the decision to be made. For example, with the digital thermometers, if hunting for potential COVID-19 outbreaks and if knowing (or suspecting) that the most probable locations are senior citizen homes and prisons, does a researcher then exclude all other readings from the database but definitely keep the addresses as part of the file to study? Probably, but then that means the researcher may not see the thousands of university students on spring break clustered on beaches that may have momentarily created their own hot spots of fever equal in number to that of senior citizens or prisoners.[52]

It has been argued that something new happens to information when it originates as Big Data, resulting in the old paradigm of qualitative/quantitative views proving less relevant. In its place, a new tension emerges, one "between the empires of raw numbers, the algorithms of mechanical filtering, and the dictates of subjective judgment," all unfolding in the question raised over a half century ago: "What counts and what doesn't count?"[53] Use of statistics presents its own issues, notably the more raw data one has, the greater the expectation that new trends otherwise hidden may eagerly be sought out, perhaps too easily or too enthusiastically. Such behaviors make "fishing expeditions" too tempting to avoid, the hunt for patterns irresistible. Given a large enough data set, some will appear that may be useful, others not.[54] An example: Random events suggest chaos reigns in the universe, in space specifically, but if scientists gather enough data about what is happening there, they begin to see an order, a logic emerging, a pattern of behavior that undermines earlier perceptions that chaos rules the universe. Given enough data about stock market transactions over a long enough period of time, one may find that women's hemlines do really rise and fall simultaneously with the rise and fall of stock prices. But is that a useful finding? As a group of experts on Big Data concluded, this new class of information "has not eliminated some of the major and longstanding dilemmas in the history of sciences and humanities; rather, it has redefined and amplified them."[55] Even issues remain of truth, beauty, and accuracy as displayed through novel and sometimes spectacular visualizations.

The various "V"s of Big Data, however, do increase the accuracy of predictions, as the pandemic models in 2020 demonstrated, even as they are changed to reflect the availability of more data and insights.[56] The potential for better decision-making improves in such matters as weather storms and epidemics, for example. But Big Data can have opacity, which becomes evident when dealing with the softer issues of the sociologist or historian, for instance. In such circumstances, to use the economist's term, *exogenous* (unpredicted, external, existential) events intervene, such as an earthquake or other natural disaster that disrupts economies, lives, or wars break out unexpectedly (as occurred with Israel and Gaza in October 2023), all disturbing prior elegant assumptions and analyses. We live in an age where everyone seems to want predictions of the future to such an extent that they can remove the risk of faulty decisions. However, we still live in a reality that leaves in the wake of decisions clusters of probabilities and uncertainties, not the certainties the technologists, statisticians, and hard scientists intimate can come from Big Data.[57] So the data miners shape the collection and use of Big Data to facilitate certainty, and in the process convert increasing numbers of people into aficionados of future prospects and predictions. The confidence level in predictions was raised, too, by the growing understanding of chaos in nature uncovered by the hard sciences.

Operating largely so far under the radar about Big Data is who is correlating and then making this data available to all the protagonists described in this and other chapters. Since the late twentieth century, corporations have quietly gained copyright control over thousands of academic journals and their content, databases that are routinely, widely, and conveniently used by the legal profession and law enforcement in the United States, financial institutions in many countries, and, too, the news media. These include such large enterprises as LexisNexis and Westlaw in legal affairs, and Elsevier in academic research. These enterprises are massive. By controlling vast quantities of information (i.e., publications and databases), they were able to place much of this material behind paywalls, thus charging for access to such an extent that individuals could only afford to consult these bodies of data and information if working for institutions willing to spend millions of library dollars on these resources. In other words, over time, large collections of information were being distributed to smaller groups of people than might otherwise have been the case, say, up to the early decades of post-World War II societies. These firms, too, did not focus on quality control—they left that to the publications themselves to handle. Over time, these data firms, which one authority referred to as "data cartels," also sold analyses of the data as forms of predictive analytics, such as correlating an individual's shopping behavior with where they lived and how much income they made to predict their future social and economic behavior, or assessing the odds of an individual being convicted of yet uncommitted crimes in the future.[58] All of this was—and is—being done without serious respect for one's privacy in the United States, but protected more so in Europe, thanks to a bit more aggressive regulatory behavior by the European Union. So access, quality, and commercial motivations also influence the role of Big Data, as described so far in this and subsequent chapters. The implications are potentially disturbing. By the 2010s, concerns about the privacy of one's information had become a full-blown issue in many nations.[59]

By being large providers of information, such corporations use their role as primary suppliers of information and data analytics about the use and users of such content, such that they can shape scientific and other research. If their analytics suggest—predict—that certain topics are more attractive to support, then funding organizations will issue grants, say, to some scientists but not to others, while journals keen on getting high ratings through number of citations (more analytics from the information cartels) then favor publishing some research results but not others, often enough without taking into consideration what experts in a discipline consider empirically important to pursue. As one observer of these large data providers explained, "we are putting Elsevier and similarly situated data and content providers in charge of determining how the world solves its toughest problems, from cancer to climate change."[60] Today, researchers in all academic fields normally know what their "impact factor" score is and to what degree that number is "good" or "bad," because it affects their candidacy for academic tenure, odds of obtaining research grants, and so forth.[61]

Meanwhile, to summarize further what the combination of Big Data, the use of AI, and comparisons of human predictions teach us (so far) is neatly summarized by a team consisting of a psychologist, an economist, and professor of business strategy worth quoting:

> When there is a lot of data, machine-learning algorithms will do better than humans and better than simple models. But even the simplest rules and algorithms have big advantages over human judges: they are free of noise, and they do not attempt to apply complex, usually invalid insights about the predictors.[62]

Their evidence and that of other researchers demonstrated for decades that human experts rarely did better as a group 50 percent of the time in correctly predicting an outcome, so Big Data and the wisdom of crowds represented an advance in the use of information.

HOW DIFFERENT DISCIPLINES INTERACTED WITH BIG DATA

While the discussion about features and issues related to the evolution of information is crucial to our focus, we must ask, how are these large bodies of data being used? It may seem a bit early in the history of Big Data to ask such a question, but the appropriation of any form of data (facts) has always influenced the shaping of information. But, we know some things. For one, large organizations that could afford to create and analyze Big Data were the first to use it; followed soon after by specialized enterprises, many born digital. Early adopters included national government agencies, international organizations, and multinational or dominant enterprises in most industries. This is the same general pattern of appropriation that unfolded with almost every new form of information and its facilitative information technologies over the past two hundred years.

Specific areas of interest included public administration, healthcare, education, media, actuarial studies, IoT, supply chains, transportation, computer sciences and telecommunications, manufacturing, retail, all manner of science, sports, and military organizations around the world. The use of Big Data became global, certainly by 2010. In fact, the usual leaders (UK, USA, Germany, and Russia) have possibly now been superseded by Chinese users and rivaled extensively by Israeli organizations. Already by 2010, commercial expenditures on Big Data projects, as measured by services provided to assist in these, were an estimated $100 billion and were expanding by 10 percent per year.[63] These expenditures continued to increase, at least until the pandemic of 2019–2022 slowed all manner of activities; but even then, it was clear Big Data uses expanded, as the story suggested at the start of this chapter.

Many uses mirrored earlier ones. Governments still wanted to reduce costs, improve productivity, and support innovations.[64] Healthcare providers are extensive users of Big Data analytics, especially for predicting outcomes of new medications and procedures, not to mention modeling costs of operating hospitals and administration.[65] Pharmaceutical research and administration represent another block of users. Health insurers, too, model how people's lifestyles and eating habits affect their health.[66] IoT and supply chains represented some of the earliest uses of Big Data, because these activities have long collected vast quantities of data that proved valuable.[67] Historians may someday conclude that IoT and health led the world in uses of Big Data in the early twenty-first century. There are now hundreds of documented cases of such uses.[68]

Because scale is one of the primary influences on those who use Big Data and for what, the issue remains a hidden, if ill-defined influencer on how information evolves. It should at least be recognized as one "elephant in the room." As an observer put it, "Information management is notoriously subject to problems of scale."[69] But, that was recognized as an issue early in the emergence of Big Data.[70] So, old retrieval methods are being replaced, a point made earlier. It is the same issue librarians, later scientists, and technologists, faced over the past two centuries.[71]

One student of the consequences of these two issues—scale and retrieval—conveniently cataloged circumstances that influence the shape of information in the world of Big Data. Christine L. Borgman, an information studies professor, pointed out that ever-larger teams of researchers form when pursuing Big Data because of the expense and variety of required skills. In some disciplines, Big Data had become essential to their core work by the early 2000s. She identified a second trend, "the blurring of the information between primary sources, generally viewed as unprocessed or unanalyzed data, and secondary sources that set data in context, such as papers, articles, and books."[72] So data sets are cited, much like scholars increasingly do with articles in Wikipedia. Third, as the number of Big Data sets has become available for use by the public and researchers at large, such as those hosted by government agencies, more people are using these, some well, others not so. But Borgman's point is that people are increasingly dipping into these for their research or to strengthen their perspectives. Quality varies in the outputs of such activities as, to use her phrase, the old librarian's ideal of "universal bibliographic control" diminishes, as it "slips even further away."[73]

At the same time, the availability of data sets encouraged cross-disciplinary collaboration, so information accumulated for one purpose could be applied in another, such as in public administration, then to craft economic development plans, or to identify potential "hot spots" for a pandemic based on the racial or ethnic composition of geographic regions. She warns that "making content that was created from one audience useful to another is a complex problem," because, "each field has its own vocabulary, data structures, and research priorities."[74] These three concerns—vocabulary, data structures, and research priorities—profoundly influence the shape of information. Add to her list information technology tools, as the behavior of computer developers and researchers in the hard sciences made clear in the past century. Big Data, thus, presents humanity with implications similar to earlier innovations in information and new ones important enough to at least identify some of the currently obvious ones.

The bigger emerging issue, thus, is that exercising Big Data and computing is both fostering collaboration across academic disciplines, but also, to use communications expert Pablo J. Boczkowski's words, it is "challenging intellectual divisions of labor." Intellectual boundaries are increasingly seen as artificial, and what might have worked better in earlier decades "seem quite dated now." Data abundance in combination with the attraction of social media is resulting in calls for innovation in the practice of data collection and use to interpret findings "through analytical lenses that creatively integrate what used to be quite separate intellectual communities."[75] As no surprise, experts ensconced in long-extant academic disciplines complain that their ecosystems are fracturing, as users of Big Data poach information and methods for using it from each other's intellectual turfs.[76] This concern raises fears that there will be a decline in traditional academic disciplines; already there has been a significant uptick in multidisciplinary projects, as already noted in this book.[77]

WHAT BIG DATA MEANS FOR SOCIETY

So, more data was available for more academics and data seekers in government and industry to pursue. This is more than a conversation about a wider selection of data at some statistical smorgasbord. Let two observers of media and technology, Danah Boyd and Kate Crawford, explain: "Big Data creates a radical shift in how we think about research." They further argue that it is a profound change at the levels of epistemology and ethics. Big Data makes it possible to reframe key questions about what constitutes knowledge, the process of research, how we should engage with information, and the nature and categorization of reality, staking "out new terrains of objects, methods of knowing, and definitions of social life."[78] Their argument harkens back to what librarians and scientists of the early 1900s contemplated was happening. But these two wisely cautioned that we also consider its limitations. "Claims to objectivity and accuracy are misleading," because—and I strongly agree as a historian with their observation—that "working with Big Data is still subjective, and what it quantifies does not necessarily have a closer claim on objective truth."[79] Researchers interpret data, information, sources, and context. Ultimately, these are soft skills. They also point out the obvious that can be overlooked: that "bigger data are not always better data," a point understood by information experts across disciplines for over a century.[80] Then, it is also true that when out of context, Big Data risks losing meaning, relevance, and its truth—the classic criticism leveled against extreme modelers and, specifically, some economists.

So, where does context fit in? Ask anyone a question with long and vast experience with a topic and they will probably respond with an answer based on their experience and context, and as discussed in various chapters about how the brain works, as a result of identifying patterns of behavior. Big data's absolute presentation of facts and AI's learning capabilities ensure that scientific discussions about the role of context will continue. But not all issues can be measured with data, or at least not measured well. Happiness is one, sense of community another, but so too the role of traditions, the definition of beauty or comfort, and the ever-popular "quality of life." Yet, one must make sound decisions regarding each of those difficult-to-measure and quantify issues. So, as one student of the issue stated the problem, "The basic methodology of data . . . systematically leaves out certain kinds of information."[81] Enter the need for trade-offs between hard data and the different perspectives and experiences of its users, hence the continuing relevance of context. The humanists advise using both, if for no other reason than to constrain the influences of biases. In each data set potentially lurks some intrinsic bias that must be controlled for. But so too, context has its own. As one philosopher explained this tension: "There is no single dependable, perfect way to understand or analyze the world. We need to balance our many methodologies," I would add big sets of information, "to knowingly and deliberately pit their weaknesses against each other."[82]

Suspend for now questions about ethical uses of information and potentially new digital divides between those with access to Big Data and those who do not to continue our exploration of Big Data's features. It is a socially constructed creation subject to similar influences as other forms of information. Nowhere did this seem more so than in a many-decades-long discussion about the privacy of one's personal information, which heated up with the likes of Amazon, Facebook, and Google using it for their business purposes without clear permission from their users. We return to this issue again because of its growing, indeed continuing, presence as an issue. Privacy had long been an issue simmering as databases and computing diffused around the world. We have only to recall the French conversations about information societies and Alvin Toffler's *Future Shock* of the 1970s. It was the establishment and rapid expansion of Internet-based businesses after the turn of the century that focused increased attention on the issue. The firms that threatened one's privacy were so linked that they are known collectively as FAANG (Facebook, Amazon, Apple, Netflix, and Alphabet [Google]). At first, the acronym was shorthand used in financial investment circles that surfaced in 2013 to designate a group of technology companies, but the news media quickly adopted it as the name for the apocalyptic horsemen of data invasion of modern times.

There are two issues often entwined but should not be for our discussion. The first is that they have become dominant players in their respective markets, leading government agencies on both sides of the Atlantic to be concerned over their potential antitrust and monopolistic behavior. While their business success was made possible by the use of information about consumers, the second issue—the one of concern here—relates to how and what they collect about individuals and policies to preserve the privacy of that data. Big Data is at the center of the issue, even drawing the ire of Amnesty International.[83]

So far in our discussion, emphasis lay on the scholarly disciplinary-oriented development and use of Big Data. Yet, by the 2010s, its largest users were in business, such as Google, Facebook, and Amazon. Big Data is considered the new gold, the new oil.[84] It is easy to accumulate, reproduce, and communicate to interested parties. As argued earlier, the advantages of scale and accumulated collections of data provide new insights of interest and economic benefits (values). One observer about its use argues past the public's concern about privacy of their information—which they do not own as only the gatherer does both in law and practice—to point out more substantive issues. Political scientist Dan Breznitz explained that

> the consequences of corporations owning your coded life, selling it as a commodity, and being able to use it as they see fit are at the core of the fabric of society. This is much more significant than privacy intrusion. Choices about data are choices about democracy and the kind of society we would like to live in.[85]

These realities are not lost on academics, businesses, government officials, or on social commentators and the obvious locations in society that can affect its collection and use: the public sector.

Governments became concerned about the practices of some commercial providers, fining some and in the process drawing the public's attention to these firms. Many incidents dot the recent history of the issue. In the fall of 2019, Google agreed to pay a $170 million fine and change its practices to protect children's privacy when minors accessed YouTube after the firm was accused of harvesting information about underage viewers.[86] Earlier that summer, Facebook agreed to create better oversight practices for handling users' data to enhance transparency and accountability of people's information.[87] Facebook was sued over privacy issues by state and federal officials in New York, the Northern District of California, and the District of Columbia (Washington, D.C.). While Apple and Amazon are major users of Big Data, as of this writing (2024), most had not been fined for abuses in their use of information, although various agencies were exploring their practices as part of their antitrust investigations.[88] The one and most recent exception to this statement involved the use of smartphones. In March 2024, the U.S. Department of Justice filed an antitrust suit against Apple, accusing it of monopolistic practices that made it difficult for users of its smartphones to access other smartphones and apps. The European Union that same month fined the company, with greater emphasis on information access issues. The pre-litigation actions dated back some years.

Amnesty International published a study in 2019 on the behavior of these firms that focused on their use of information: "services provided by Google and Facebook derive revenue from the accumulation and analysis of data about people. Instead of charging a fee for their products or services, these businesses require anyone who wishes to use them to give up their personal data instead."[89] It accused them of, "using algorithmic systems to analyze this vast amount of aggregated data, assign detailed profiles to individuals and groups, and predict people's interests and behavior."[90] The report spoke specifically about Big Data: "The rise of 'Big Data' and continuous tracking of people's lives online has created a 'golden age of surveillance' for states, providing authorities access to detailed information on people's activities that would have been unthinkable in the pre-digital age," while Facebook and Google were able to use such data with minimal regulatory oversight.[91] Since 2001 both governments and these companies expanded their use of such data in relatively unfettered circumstances with near impunity.

What data did these two companies collect that proved so concerning to officials and the public by 2010? These included records of people's browsing activities, search terms, websites they visited, and from what locations by monitoring, for instance, one's use of the "Like" function on Facebook. In 2018, Facebook reported that the "Like" button appeared on 8.4 million websites, the "Share" button on 931,000 websites, with each instance tracked by Facebook. Both Google and Facebook collected names and addresses of e-mail recipients, location records, and timestamps on these and photographs.[92] These data were used to predict the behavior and interests of individuals, which they sold to advertisers and "app" producers. It did not help their cause that in 2010 Eric Schmidt, Google's CEO, was ominously quoted saying, "We know where you are. We know where you've been. We can more or less know what you're thinking about." To be perfectly accurate, he prefaced this comment by saying, "With your permission you give us more information about you, about your friends, and we can improve the quality of our searches."[93]

Media, then academic researchers, began probing these uses of Big Data in the 2010s. For example, Shoshana Zuboff, professor at the Harvard Business School and a highly regarded expert on the combined role of human behavior, economics, and information technologies, co-authored with James Maxim a lengthy criticism of these companies and how they used Big Data, *The Age of Surveillance Capitalism: The Fight for a Human Future at the New Frontier of Power* (2019).[94] Examining briefly their comments conveniently flags the growing concern with privacy issues that already had attracted the attention of public officials in Europe and the United States and that must have influenced such extensive users of information as scientists. Using their phrase "surveillance capitalism," they argued that these companies generated profits from selling information about people and that there were inadequate protections for such data. Zuboff, in particular, feared for people's autonomy to act and the threat these data posed for democracy's health. Byproducts of such realities include the dissemination of fake news aimed at specific audiences, surveillance of people's political behavior (a major concern regarding China, Russia, and Hong Kong), and destabilization of democracy and capitalism.

The *New York Times*, with a history of concern over the negatives of Big Data, published a lengthy article in January 2020 by Zuboff that exposed her ideas to its six million subscribers, titled "You Are Now Remotely Controlled."[95] Several of Zuboff's comments capture the essence of her concerns: "we celebrated the new digital services as free, but now we see that the surveillance capitalists behind those services regard us as the free commodity. We thought that we search Google, but now we understand that Google searches us." Additionally, "privacy policies are actually surveillance policies." The expanded recent use of facial recognition technologies posed another new assault on one's privacy, with her calling "privacy as private" a "delusion." The new capitalism "signals a power shift from the ownership of the means of production" to "ownership of the production of meaning." This development raised fundamental questions about knowledge and power in contemporary society: "Who knows? Who decides who knows? Who decides who decides who knows?" Zuboff blamed Google, Facebook, Amazon, and Microsoft for driving these changes underway.[96]

As I was writing this book in 2021–2024, the debate about the role of information in society had reengaged on the battlefield of electrified information. The years between the 1980s and the early 2010s now appear to have been, in comparison a quiet hiatus, while technology diffused and the quantity of digitized data increased, seeping into societies around the world in massive, if in not fully evident, ways. We have yet to see the full suite of consequences revealed for societies at large. But already, a new, interesting, possibly profoundly important development had been simultaneously bubbling in academic circles that one should acknowledge, as it may already be having a profound effect on how science is done, even if it is too early to proffer a projection.

Briefly, do scientists need theory anymore now that they work in a data-driven environment? For half a millennium, it was canonical that scientists sought explanations of how natural phenomena functioned, with these articulated as theories that could be tested, validated, and used to explain physical realities. The approach worked well. The use of theories became a form of intellectual economy, the brief explanation for how something works without having to repeat each time observations, replacing documentation of individual occurrences with general rules that apply to similar phenomena. Would Big Data be better at describing a phenomenon through the crush of vast quantities of data? As a group of researchers who posed the question of implications put it, "What new conceptions of theories, and of science for that matter, do we need in order to accommodate the changes brought about by Big Data?"[97] That question raises others concerning implications for theory-based science, for relationships between science and society, and for what kind of information is lost in the process.

What would René Descartes (1596–1650), Isaac Newton (1643–1727), or Stephen Hawking (1942–2018) have thought? The gravity of such questions was worthy of their consideration. Why? As a group of scientists explained: "The vast quantity of data produced on social media sites, the ambiguity surrounding data ownership and privacy, and the nearly ubiquitous participation in these sites make them an ideal focus for studies of Big Data."[98] Like Descartes, Newton, and Hawking, today's scientists do not work absent influences of their day, and today that is the growing universe of Big Data. I already explored in earlier research the changing—actually forced collaboration—of pre-Big Data disciplines that had previously been carefully built up with their own infrastructures of language, ideas, beliefs, societies, departments, and publications.[99] But users of Big Data did not respect those boundaries to the extent exhibited by earlier generations and were now threatening these.

In 2008 Chris Anderson, then Editor-in-Chief of *Wired* magazine, made such a startling comment that even distinguished scientists gathering at the Aspen Institute could not ignore it: "the data deluge makes the scientific method obsolete."[100] He thought having more taxonomies and models was no longer as sufficient as sifting through data in novel ways to identify meaningful correlations, an approach "that requires us to lose the tether of data as something that can be visualized in its totality. It forces us to view data mathematically first and establish a context for it later."[101] He accused physicists and geneticists of having strayed into a land of speculative theorizing because their practitioners did not possess adequate testing models. Examine enough data and "Correlation is enough. We can stop looking for models. We can analyze the data without hypotheses about what it might show." He added, "We can throw the numbers into the biggest computing clusters the world has ever seen and let statistical algorithms find patterns where science cannot."[102]

The room, filled with scientists listening to him, reacted predictably: Anderson's polemic went too far. As one responded, "You have to have some basis for asking questions."[103] In a separate publication, another scientist later stipulated that "Correlations are a way of catching a scientist's attention, but the models and mechanisms that explain them are how we make the predictions that not only advance science, but generate practical applications."[104] Theory helps to validate predictions; predictions then endow theory with confidence. Another long-embraced defense held that "Why does deduction work? Well, because you can prove it works. Why does induction work? Well, it's always worked in the past."[105] Others believed that data's correlations tested theory and improved them. Does the practice of formulating theories need to evolve in the face of so much Big Data? John Seely Brown thought so: "In some ways, the more data you have, the more basis you have for deciding that something is an outlier. You have more confidence in deciding what to knock out of the data set—at least, under the Bayesian and correlational-type theories of the moment."[106] He opined that Big Data could cast light on "generators" that lead to new theories. Shannon's "noise," and what Google, Amazon, and others thought in the beginning was "data exhaust"[107] could point to new ways to develop theories and ontologies.[108]

These issues concerning the role of theory are examples of the debates regarding canonical beliefs. In the humanities and in earlier times in other disciplines now branded as sciences, members never fully embraced the centrality of theory, such as in history or librarianship, or among earlier sociologists and political scientists. We will have to suspend further debate about this matter as it is just barely being considered and leave it to future students of information to explain how it all turned out. In the meanwhile, scientists and other observers remain both optimistic about the usefulness of Big Data, if also nervous, "because data technologies are becoming so pervasive, intrusive and difficult to understand. How shall society protect itself against those who would misuse or abuse large databases?"[109]

CONCLUSIONS

What are we to make of Big Data's increasing role in such activities as IoT with so many sensors joining AI to direct activities, including those of humans? How should one respond to the manipulation of our tastes by online vendors, political parties, and purveyors of news and information slanted in support of some social or political agenda? Jürgen Renn, director of the Max Planck Institute for the History of Science in Berlin, has long been a thoughtful student of how scientific knowledge evolves. He, too, sees Big Data as both a continuation of earlier knowledge traditions, but also as something new. It is a social web energized by a new force that "has increasingly become a means of acquiring and transforming control over societal and economic processes, giving Big Data 'control'."[110] *Control* is not a word one used so much to describe the power of information in recent decades. Before, it was seen more as a tool to *inform*, even in the 1980s and 1990s (pre-Internet) when "expert systems" were supposedly altering human behavior. Let him explain the changeover: "the information-data-information cycle of the pre-Web market economy is thus effectively changing into a 'data-information-more data' cycle enabled by the Internet and controlled by those owning the means and infrastructure for accumulated and using that data—be they private companies or the state."[111] In his way, Renn joins the scientists discussed above in their belief that somehow Big Data has features that cause information to change. Larry Page, CEO and co-founder of Google, understood how it fit with computing when he exclaimed, "sensors are really cheap," and those are what produce so much data, indeed one could argue, to the extent that it has become impossible not to collect it.[112] His world became what Zuboff was describing with a more passionate narrative.

Let this chapter end by clearing up a potential confusion: the meaning of metadata versus Big Data because it is appearing more frequently in today's discourses about information. Had I been asked to differentiate between the two terms in the late 1990s or even early 2000s, it would have been difficult to do, but less so now. The notion of metadata is credited to David Griffel and Stuart McIntosh at MIT in 1967, when they noted that "we also have statements in meta language describing the data relationships and transformation, and ought/is relations between norm and data."[113] It is normally described, correctly, as "data about data," with "meta" suggesting that there exists information that sits above some data, providing information about other information that can be used to track data, organize it, and most importantly to describe it. If this vaguely echoes what librarians were thinking when they put together bibliographies, annotated lists, and summary 3 x 5 cards, then you would be right. One could devote an entire chapter to discussing metadata, but this more than fifty-year-old concept seems antiquated. Since not all data is Big Data, the software, the tools, and the organization of databases into groups of other databases remain widely in use. So metadata continues with its syntax, schemata, standards, and uses co-existing side-by-side in many organizations with Big Data.[114] Walk into a data center that has metadata activity, and one bumps into "data warehousing," the repository for information formatted and controlled in late twentieth-century ways. In fact, walk into any large data center and you will see metadata alive and well both in the building and in the software housed there and out in the Internet.

This chapter began with an example of Big Data used to help explain to the public what was occurring with the pandemic. I close with a revisit. A few reports remind us of the revelatory nature of Big Data:

- An examination of the top Google searches in the United States related to COVID-19 by topic between January and April 2020 uncovered how many queries were made, identified regional patterns, and what questions

were asked during each phase of the pandemic, from when the first death was reported to when the American government released its first economic stimulus package. Each period had different questions from the beginning (basic knowledge, symptoms, treatment) to next government responses and how to apply for funding, to economic impact, and so forth.[115]
- Within a month of entire nations going into quarantine, hence less travel, government agencies were reporting rapid declines in CO_2 emissions around the world, complemented with satellite pictures of cloudless skies in India and statistics on changes in the West.[116]
- Using cell phone data, within days of American states lifting stay-at-home orders, communications providers could report that 25 million Americans were traveling, accompanied by data and graphics to explain where, how many miles, and when.[117]
- Internet traffic changed during the 2020 spring lockdown, and that data was reported out quickly, showing how much traffic went to Facebook, Netflix, YouTube, and other sites and for what content. Phone usage declined, and that too was reported.

All these reports were based on many billions (possibly trillions) of data points captured initially by sensors and managed by software, and extracted rapidly through human queries. These were presented to us mortals with graphics, not just statistics. Meanwhile, South Korea was receiving a great deal of positive attention because it was using smartphone apps to track coronavirus contacts.[118] The *New York Times* reported where wealthy New Yorkers moved to escape the pandemic by tapping into the U.S. Postal Service's data on change of address requests.[119] That is not to say older statistical databases were gone; far from it, because, again citing *The New York Times*, the number of new COVID-19 cases that appeared each day, statistics on the number of deaths, and where all the cases existed represented more traditional nose counts from the CDC, Johns Hopkins University, and local media.[120] It is difficult to imagine that the public will retire to older, more narrative and statistical reporting after what they saw regarding COVID-19. The Internet and Big Data opened a new world to many hundreds of millions of people.

But that same door also let in all manner of inaccurate, misleading information that struck at the heart of the veracity of facts—what many scholars were beginning to call "noise."[121] It is to the broader issue of faulty data and its influence on the quality of information to which we turn next.

Chapter 5

How Factual Is Information and Why Should We Care?

All control, in essence, is about who controls the truth.

—Joseph Rain[1]

Other ways to title this chapter might be to ask: How many fake facts are there? Do we live in factually fragile times? On April 14, 2020, the *Washington Post* reported that President Donald J. Trump had made over 19,000 false or misleading comments over the course of 1,170 days that fact checkers had disproven.[2] *Wikipedia* even maintains an entry to catalog his false statements, including charts and graphs tracking his pronouncements over time by the news media. The entry had 259 endnotes on the day I checked it. It appeared that this source was being updated multiple times each week.[3] Lest any politically liberal or smug academic shake a finger at me and says "tsk, tsk," a reputable tracking organization that does analytical studies of scholarly publications is famous for announcing how many "dubious" academic journals are published. In 2020, it reported that about 1,000 "predatory journals" had publications clearly not scholarly just in one year: 2010. These emerged largely in the hard sciences and health; but ten years later, the number had climbed to 13,000. Many authors with weak academic credentials or fake or trivial scholarship appear in these journals, including some legitimate (if gullible) scholars. These journals are published all over the world as open-source publications.[4]

All of this is before or in addition to any discussion one could hold about Russian meddling in American and European elections since 2014 through Russia's Internet Research Agency's publication of many thousands more fake statements passed through many tens (or possibly hundreds) of thousands of fake Internet sites. This is in further addition to the fake conspiracy stories about President Obama's birthplace or religious affiliation, or the seemingly endless round of COVID-19 stories about cures. Then there were the "facts" circulating about the American 2020 presidential election being corrupt that led to the January 6, 2021 deadly assault on the U.S. Capitol. Fake facts have consequences; in that latter incident 5 people died, some 1,000 were charged with criminal law violations and most were serving prison sentences. The United States remains badly split on political and social issues to an extent not seen since the days of the Vietnam War a half century earlier. Europeans are not immune from fake facts and the war in Ukraine has served up both fake facts and, to use an old-fashioned term, propaganda, both originating from all over Europe, North America, and Asia, and not just from Russia or Ukraine. As this book was being published in late 2024, the truth trackers were still documenting fake facts uttered by Trump, now in his third presidential campaign, as they had for nearly a decade.

One would think everyone produces lies, especially more so, perhaps, since 2008. They would be right to conclude that we live in a Golden Age of Fake Facts, which has a stronger claim to being a more accurate name for our time than the Age of the Potato. But the issue is, of course, far bigger and more widespread than any discourse one could mount, say, about Big Data. The quality, meaning accuracy, of information has become a topic of intense interest. It is a pervasive issue for our time, which is why we need to understand it. Two students of how libraries, online media, and Google (in particular) observed that while librarians and social media firms were either creating quality information or encountering fake facts (e.g., from Facebook), the public was having to function in an increasingly fractured information ecosystem. Let them explain: "The filter bubble became all too real. Conspiracy theories competed with reality. Authority collapsed and institutions were threatened. Validated information sourced from publications and the expertise that they held dwindled in social significance," despite efforts of online information providers and libraries to do otherwise.[5] The Pandemic during 2020–2023 forced more people to rely on information from a broader information ecosystem than before, with online sources becoming more widely available.

To pick up on an observation of the experts just quoted, increased use of possible fake facts raises an ugly thought: "It is not clear that new generations will be able to distinguish trustworthy knowledge from misinformation."[6] The recent surge in new artificial intelligence (AI) tools since 2022 has made this problem even more urgent, if not intractable for the time being. Thus, such concerns justify the need to dedicate an entire chapter to this emerging form of changed information. But, it gets worse.

An expert on information security, Thomas Rid, focused more specifically on the role of espionage and how the Soviets and Americans used information and false facts during the Cold War, reminding us that the topic has a longer history that predates our current circumstance. He turns to more philosophical means to consider the issue of truth that informs the subject of both this and our other chapters. He argues that there is a "cultural tension within truth itself," that is to say, between two widely held understandings of the concept of truth that "stand in permanent opposition to each other." The first is "positive and analytical; something is true when it is accurate and objective," what students of information discuss the most—science-based researched, relies on observation, and so forth. It focuses on today and the future. The second "corresponds to belief, not facts. Something is true when it is right, when backed up by gospel, or rooted in scripture, anchored in ideology, when it lines up with values." His ideas harken to an earlier time or anticipate a new one.[7] Both types are exaggerated, idealistic, and clichéd, nonetheless, he observed, were present in the world of spies and Cold War behaviors of government officials. So they were real, caused action, and cost lives. Such actions codified how to create and promote fake facts. This behavior continues today, for example, in how Russia uses the Internet to defend its invasions of Ukraine in 2014 and again in 2022.[8]

I proceed in a three-step manner. First, it must be understood that fake facts have circulated for centuries; here, I describe some of their features, using again the United States as the teaching case. Then, the special role of computing and, more precisely, the Internet to analysis, because just as these technologies facilitated the creation and distribution of useful information in many disciplines and organizations, they also did the same with inaccurate and mendacious data. A third discussion involves the existence of a new form of information illiteracy exacerbated through the massive use of the Internet by the public. This chapter ends with recommendations about what to do, in the belief that it is possible to reduce the currency of inaccurate or dishonest information to restore a more balanced allegiance to the kinds of information available in the twenty-first century. New skills needed in the past were embraced; it is time for another round of aligning people's capabilities with the bodies of information shaped by events of the past two centuries.

LONG-TERM PATTERNS OF FALSEHOODS

Since the eighteenth century, three types of incorrect or fake information have circulated in North America. Many were misrepresentations of facts swirling about during national political discourses or as a routine part of local and national elections. A second type involved misleading or outright falsehoods about a product's features, with messages usually distributed through advertising, occasionally misleading "scientific" reports, but always purposefully promulgated by the vendors of such products and services. The third was issue-based falsehoods or misrepresentations of the truth, a group of which became widely evident following World War II but were already in evidence at the dawn of the twentieth century. Each existed across American society; no state or city was immune from exposure to these, no generation either, nor any age or racial group. Much of what appeared was intended to mislead the public or to influence business and government officials. Overlaying each of these types of mistruths was a heavy layer of opining by supposed "thought leaders" (a late twentieth-century appellation) or individuals at large. Examples of the latter are people's blogs, postings on Facebook, and Twitter sound bites. All three do not require credentialing to weed out those not expert in a topic.

Besides the obvious problem that fake facts can cause someone to take a supposed medicine that ends up killing them or voting into office someone whose values and policies do not match one's expectations, there are the motivations of their promoters and the consequences. The distinguished philosopher and retired professor from Princeton University, Harry G. Frankfurt, exclaimed,

> that bullshitters, although they represent themselves as being engaged simply in conveying information, are not engaged in that enterprise at all. Instead, and most essentially, they are fakers and phonies who are attempting by what they say to manipulate the opinions and attitudes of those to whom they speak. What they care about primarily, therefore, is whether what they say is *effective* in accomplishing this manipulation [and so are] indifferent to whether what they say is true or whether it is false.[9]

Let him state the obvious:

> higher levels of civilization must depend even more heavily on a conscientious respect for the importance of honesty and clarity in reporting the facts, and on a stubborn concern for accuracy in determining what the facts are. The natural and the social sciences, as well as the conduct of public affairs, surely cannot prosper except insofar as they carefully maintain this respect and this concern.[10]

We do not need to engage in the torrid and extensive debates about relative truths or the ideals of the Postmodernists; leave that to the philosophers and to other "bullshitters" to use the distinguished professor's term.[11] We need to further understand the kinds of false facts that have long existed in American society.

The first type—political untruths—has circulated since the dawn of the United States. Much of it can be characterized as slander impugning the views and reputation of candidates.[12] Thomas Jefferson in 1800 was accused of sleeping with his female slaves, Andrew Jackson's mother of being a prostitute and his wife a bigamist, Abraham Lincoln of being "dumb as a monkey," and John F. Kennedy of only doing the bidding of the Pope contrary to national interests.[13] President Trump is famous for giving people nicknames to imply that they embraced views or behaviors divergent from those of supporters he sought to impress: "Sleepy Joe" for former vice president Joe Biden, "Low Energy Jeb" for Governor Jeb Bush, and perhaps the best known, "Pocahontas" for Senator Elizabeth Warren.[14]

The second—misrepresentation of facts regarding a product, usually false advertising—is memorialized in American history through quack medicines, snake oil tonics, and exotic cures for cancer in Mexico, of course, a country which has long been blamed by conspiracists and others as the source of many mysterious or bad things. There were essentially no regulations to control dissemination of such false facts until early in the twentieth century when Congress established the U.S. Food and Drug Administration to address part of the problem. But even then, false advertising of medicines, packaged foods, and the quality of various meats continued appearing throughout the twentieth century.[15] The latest iterations involved vitamin and health supplements, although they, too, had a history dating to the early nineteenth century.[16] Such messages originated with individual vendors (manufacturers), often compounded by traveling salesmen in the 1800s, later by exaggerated advertising (e.g., those put on television), and in magazines in the 1950s such as by Camel Cigarettes touting its products as safe and the favorite of medical doctors.[17]

The third—issue-based misrepresentations—were the propaganda, press releases, pseudoscientific reports, and others produced by lobbying or front organizations promoting specific points of view. These became most widely widespread after World War II, but had become an unidentified genre by the 1920s.[18] Famous examples include the tobacco industry's campaign to block the control or elimination of cigarettes from the 1950s to the end of the century, when a number of the industry's major vendors lost court cases accusing them of hiding the fact that their products caused cancer. However, these and other firms continued to use the same tactics to block regulations of e-cigarettes.[19] A second instance involved the petroleum industry (and others) contesting the validity of climate change and the need to move to non-petroleum alternative energy. It is an ongoing battle in the United States, the only country whose national government declined to improve environmental conditions until the start of the Biden Administration in 2021.

These, and others, share common features. They are the result of coordinated and funded activities by major enterprises within one or more industries. These are well-lubricated with the advertising, public relations, lobbying, expert testimony, and publications needed to mount sustained national campaigns extending over many years. They have their own publishers, websites, blogs, academics appearing in journals and books, "think tanks," and industry associations with innocuous names to mask their purposes. They cluster in major public sector cities such as Washington, D.C., Brussels, London, Berlin, and Paris. Members of these organizations and companies also move back and forth from industry to government and think tanks then back to the private sector, most frequently in the United States.[20]

There is yet another way to characterize these various shades of truth, false statements, and lies, and that is by type. Rumors are the most obvious, such as those about slave revolts in the 1700s and 1800s, job layoffs in the 1990s and 2010s, or Elvis Presley sightings in American supermarkets since the 1970s. Conspiracies are often the largest in that they engage more people than rumors and circulate for decades. Examples include the causes of wars, who assassinated various presidents, sources of pandemics, and bad food. Old lies were always in evidence, such as denying (or believing) that aliens landed in Roswell, New Mexico in the 1940s, or questioning whether the Moon landing even took place in 1969, earlier claims that Spain blew up the USS *Maine* in Havana (Truth: Spain did not), Russians did or did not intervene in each American state and national election from 2008 through 2024, and so forth. Religious forecasts have long been widespread, such as the end of the world was coming (1840s), visitations of the Virgin Mary in Wisconsin (1859), or that one of the Kennedy brothers would come back in Dallas (2022) come quickly to mind. Add End of Time statements circulating in 2020 during the pandemic and subsequent racial protests.[21] Most recently, the evils of COVID-19 vaccines replicated similar stories since the origins of smallpox vaccines in the 1700s. To prevent massive outbreaks in the Continental Army, General George Washington had to mandate that his troops be inoculated. It seems some things never change. President Joe Biden had to do the same with Federal and U.S. military personnel regarding COVID-19 vaccines thirteen decades later.

Military falsehoods circulated in every war in American history, such as convincing the Germans that the Allies were going to invade Europe somewhere other than at Normandy in 1944. Political scandals were a growth industry of falsehoods in every year of American history and, of course, swirled around every president and most governors. Politicians participated, such as President Richard Nixon denying involvement in the Watergate cover-up (1970s), President William Clinton stating "I did not have sex with that woman" in the 1990s, and then, President Trump's pronouncements.[22] Some of the most famous fake news stories involved the announcement of Mark Twain's death, which he famously denied had occurred (1897); or reports from comedian Johnny Carson (1973) that the nation was running out of toilet paper, causing a shortage as people rushed to acquire some; or that

Martians had invaded Earth, reported by Orson Welles in a radio broadcast in 1938.[23] In each instance, many people took as truth what they read and heard.

Other typologies are emerging that should be recognized as possibly useful as we come to learn more about the role of fake facts. Claire Wardle, an experienced expert on journalism and social media, argues that there is a great deal of "information pollution" that

> conflates two notions: misinformation and disinformation; the former is false but distributed in the belief that it is true—what a friend or uninformed relative might post on Facebook—while the latter is known to be false but is distributed on purpose to mislead others—what spies and political foes might do.[24]

She sees a range of fake facts from misinformation to disinformation to malinformation. The first might be satire or errors, the second fodder to feed conspiratorial theories or to spread rumors, the third deliberate, such as for revenge, or to alter a context. She categorizes the combination of the three as "information disorder," and catalogs seven features: satire and parody, false connection, misleading content, false context, imposter content, manipulated content, and fabricated content.[25] While we will not engage in a discussion about her typology, it is further evidence that inaccurate information has an ecology and world of its own, one that assists in justifying the need for both this chapter and for further exploration of its role in explaining how information changed over time.

Across the sweep of at least American history, people behaved with respect to how they accepted and used facts in at least five consistent ways, regardless of the veracity of the information.[26] The intensity of these practices ebbed and flowed over time, depending on the topics. These were in evidence in such far-ranging examples as how people responded to stories and rumors of presidential assassinations and outrageous advertising claims to political shibboleths and falsehoods. These episodes help us to understand how debunking occurred over the past two centuries, a topic not explored in great detail here, except when discussing the problem of information illiteracy below.

The first and most fundamental behavior observable across all manner of information use is that everyone involved, that is to say, those actively engaged with an issue, appropriated accurate and inaccurate information as primary tools for achieving their goals. Scholars did it, politicians certainly did so, advertisers and public relations experts built their careers on this, and the public sought out and shaped their views based on all manner of information. Readers will quickly recognize that perhaps the most obvious and pervasive demonstration of such behavior occurs in the extreme in U.S. presidential elections. Jaded presentations of the truth in support of one's position on an issue have been common practice since at least the eighteenth century. Conspiratorial discussions represent a clear example of the same behavior. It seemed every political and ideological faction of their time presented evidence of myriad groups assassinating presidents. In Kennedy's case, that meant accusing the FBI, CIA, Department of State, President Johnson, Fidel Castro, the Soviets, anti-Castro Cubans, Texas politicians, the oil industry, disaffected Floridians and others in Louisiana, Mexicans and also Mexico, organized labor, and the Mafia.[27] In Lincoln's case, it was disaffected Confederates operating out of Canada, officials of the Confederate Government in Richmond, Virginia, and John Wilkes Booth working with a small cabal; but in the end, Wilkes did it. This behavior occurred while there was also growing literacy, people turning to printed facts for information, and a growing willingness to consider fact-laced arguments and points of view, all applied to appropriating fake and factual information. The good and the bad co-mingled and circulated. Then, as now, therefore, discriminating between truthful and misleading or false information proved problematic for consumers of such material.

A second behavior was the presentation of information of all types and degrees of accuracy in the literary and rhetorical style of their day. In the 1820s, Jackson's wife was accused of being an adulteress; it would be more difficult to imagine someone making such an accusation in the 1930s, 1980s, or 2010s of a presidential spouse. Lincoln was called a "monkey," and not until Trump became a political figure was such crude language used again. For decades in the nineteenth century, such personal insults were normal, far less so later. Statements without qualification resting on facts and other evidence ebbed and flowed based on the technologies used. Newspapers in the 1800s carried longer stories than could (or did) radio or TV, until the arrival of cable news talk shows late in the twentieth century.

Vetting reflected the standards of their time. For example, lacking scientific data in the 1700s meant opinions had to be buttressed with biblical notions or folk wisdom. Today, numeric data and charts and graphs often support a point of view. With the arrival of experts in the mid-nineteenth century, they were trotted out with their credentials and a growing supply of facts in support of one perspective or another. One did not have to be a medical doctor in the 1810s to opine on a medical issue, nor even in the 1880s, but by the 1920s, one better have had M.D. after their name to be taken seriously. Doctors were recruited to sell cigarettes in the 1950s, while in the 1990s professors were recruited to debunk global warming. To be credible today, it helps to have published a book on what one is opining about in the media, but with blogging and Facebooking, we are back to practices in evidence prior to the American Civil War.

A third practice is that individuals and all manner of organizations weaponize information to achieve their objectives. The use of the word *weaponized* associated with information is a new construct. It originated as an American military term during the Cold War to describe turning specific technologies into weapons, such as rockets. In the early years of the twenty-first century, other issues were weaponized, such as femininity, religion, ideology, and facts. After 2012, it seemed every political issue was being weaponized along with many Internet behaviors, such as online harassment or WikiLeaks against political foes.[28] The meaning is obvious: to adapt for use as a weapon. It is now a fashionable term.

In earlier decades, one would have conceived of information as a tool to be used for some practical purpose, not solely to conduct war or to mislead others, although that was done, too. Weaponizing information involves shaping facts or fiction to persuade someone to their point of view, to mislead, to incite discord, or to cause people to take action. The activity is purposeful, such as when East Coast editors in the 1890s encouraged their readers to support Cuban rebels against their Spanish rulers, or in the 1920s to make the public nervous that anarchists and communists posed a serious danger to Americans, or President Trump wishing to create the impression that Democrats were socialists bent on upturning American democracy and that the 2020 elections were "rigged." Patent medicine and tobacco manufacturers were notorious weaponizers of information, climate change lobbyists, too, if more modestly.[29] Creating images for political candidates was elevated to a high art in post-World War I America, largely at the national level but frequently by governors and state legislators, too. John F. Kennedy was presented as healthy when, in fact, he was not. Richard Nixon was "Tricky Dick" but today is viewed as more moderate than in the 1950s and 1960s.[30] Hillary Clinton depicted Trump's supporters as the "deplorables," while Republicans enjoyed the nicknames their presidential candidate bestowed on rivals, such as "Little Marco" Rubio and, it seems the favorite of 2016–2019, "Pocahontas." Each damaged the images of a rival.

A fourth practice that came increasingly into use after the establishment of large organizations, such as corporations, national political parties, and lobbying think tanks and associations, was the creation and use of fake facts and misinformation to promote their causes. This is the practice of scaling information. Prior to the 1880s, most creators and purveyors of misinformation were individuals, small groups, or newspaper editors. These never went away; in fact, they came back with a vengeance once individuals could present their views through blogs and websites on the Internet.[31] Unions and national industry associations leveraged their size and reach to shape information in support of their organizations' interests. For example, unions exaggerated stories about working conditions in defense of their demands for new working conditions. The tobacco industry traded in boldfaced lies. The U.S. Department of Defense misled the public about how unsuccessful it was in waging the Vietnam War in the late 1960s and early 1970s. Medical advocacy groups routinely agitated for funding their pet disease research agendas, while climate change advocates exaggerated the harm caused by factory emissions.[32] As early as the 1830s, Southern politicians minimized the importance of slavery when advocating for American dominance of the Caribbean, an area threatened by abolition movements supported by Great Britain. Historians believe the Monroe Doctrine, promulgated in 1823, had less to do with keeping European empire builders out of Latin America and more to protect a large region for slavery and U.S. economic hegemony. But the polemics around the concept remained extensively debated for generations.[33]

One additional fifth practice was in evidence largely after the 1960s: individuals had less impact in the creation and flow of misinformation, although they never went away. After the arrival of social media with the Internet, they often did their communicating under the umbrellas of websites and associations. This trend reflected the ebb and flow of shifting sources of influence underway since the late eighteenth century. As outlets for misinformation came under the control of organizations and as other associations professionalized disciplines with one of their objectives to block the creation and flow of false facts, the influence of individuals waned. There were exceptions. Leaders at the National Rifle Association (NRA) were able to use such organizations to project their perspectives.[34] Celebrity newspaper editors could do the same, as could their outlets, such as cable news promoting one perspective over another, each accusing the other of spreading falsehoods, such as Rachel Maddow at MSNBC with liberal views critical of the political right and Laura Ingram at Fox News with conservative perspectives opposing the political left, both TV personalities in post-2010 America.

Before and after the arrival of the Internet, people promoted their views and facts with varying degrees of capability, often to audiences not always able to scrutinize what they were being presented. That is why this chapter ends with a discussion of the issue. Often all five practices complicated matters by masking sources of information, even regarding who was putting it together. When I, and a colleague, William Aspray, studied fake news and facts, we ran into this problem. It did not matter if we were looking at the 1820s, 1920s, or today. We had to write an entire book just to explain the scrutinizing and fake facts problem involving the Internet. To keep from having our research sprawl out of control, we stopped our investigation with events through 2015, leaving to others to explore the election of 2016 and beyond.[35] Hundreds of thousands of misleading websites had led to the creation of others that did fact checking; so all is not "bleak," to use Peter Burke's word, an expert on ignorance and fake news.[36] All of that was even before one could discuss the insertion of falsehoods into American society by Russian and Chinese agencies and their allies.³⁷ Rough estimates suggest that up to 60 percent of what appears on the Internet is now faulty facts.³⁸

ROLE OF COMPUTING AND THE INTERNET IN THE CREATION AND DISSEMINATION OF FAKE FACTS

If computing facilitated the further creation and dissemination of accurate information on a large scale, did it not also do so for fake facts? This is a different question than the statement just made about how much inaccurate material shelters on the Internet. In prior chapters and earlier volumes of our study we learned that how people shaped information was in part a function of the technologies (i.e., platforms) they used, such as printing presses, the format of books and magazine articles, terse commentaries on radio and television, the assemblage of massive files, and quick access to these made possible by online systems, and now the Internet. Recall that computing came in essentially three ways.

The first involved the use of large computer systems, beginning in the 1950s and continuing to the present. Because these were (and are) expensive and complicated to operate, they were and continue to be controlled by large organizations, corporations, and government agencies. These systems focused on serving their internal users, so the issue of false facts circulating within these systems was of less concern outside their owners' organizations. When these circulated, it was less due to malicious intent than to faulty research and development or to normal errors encountered in engineering or programming. Such organizations were not motivated to produce bad data, more only to conceal some information from the public, as occurred with the Pentagon Papers published in 1971 that revealed American activities during the Vietnam War.³⁹ These observations apply to minicomputers that proliferated large and medium-sized organizations under similar circumstances as for large mainframes.

The second diffusion of computing involved personal computers. That technological innovation represented a major step change in computing because now individuals could create, use, and disseminate information *on their own*. They did not need to have their content reviewed by corporate lawyers or public affairs (i.e., communications departments) unless they were speaking on behalf of their employers. Tens of millions of people in the 1980s could opine and publish whatever they wanted from the comfort of their homes without having their materials vetted by experts. Thus, by the 1990s, and just before the wide availability of the Internet, millions of people had quickly become accustomed to sharing their views on all manner of topics with friends, colleagues, and in pre-Internet chat rooms. Yet even they did not participate extensively in the propagation of fake news. That activity remained the purview of lobbyists, issue-based associations, and other institutional and political players.[40] However, PCs became the initial technology onramp to the information highway of the Internet in the 1990s. In other words, to use the Internet, one either had to access it through their terminal at work or PC at home; doing something provocative using one's work terminal (or PC) put their job at risk, but the one at home, or that which was the property of a malicious group, was another matter.

The Internet led to many changes in the distribution of false and misleading information, to the point that more than half of what appears there may be bogus. Most importantly, the fact-checking vetting practices of book, magazine, and newspaper publishers, while these continued, could be bypassed by individuals less concerned with such matters and, by the end of the first decade of the new century, by software that created fake websites and questionable content. Yellow journalism of the 1890s came back, with algorithms (programmed software-based instructions) creating and spreading messages that promoted such issues on the American political right as gun rights, false information about laws, racism, and white supremacy issues. From the political and social left, much appeared that was misleading about climate change and the evils of the political and social right, while both sides co-opted materials they found on the Internet that bolstered their points of view without vetting. This last practice became widespread during the national elections in 2016, 2020, and 2024. Meanwhile, the number of professional reporters who could counteract such information shrank.[41]

Three behaviors made possible by the Internet became evident by the early 2000s. First, social media platforms collected content from the Internet as news aggregators, presenting these to individuals whose prior interests and searches indicated to software an interest in the topic, regardless of the veracity of the material. If one did searches on, say, eating habits of chickens, Facebook and Google would make sure that person was exposed to other postings about chicken culinary habits, regardless of whether they were true or not. Algorithms are agnostic on such matters. So, those posting faulty or misleading content stand a good chance of their materials going to a wide audience, that is to say, to many people interested in what chickens eat, especially if they paid to have their postings ranked higher in Google's search algorithms.[42]

Second, people distribute stories and postings that arrive at, say, their Facebook page that they like, which they think their friends and relatives would like too, and so pass these on to them without vetting. Readers may have relatives who do this, especially during national political campaigns. President Trump was famous for retweeting messages to his followers.[43] One study of his tweets counted 217 retweets since he became president through November 2, 2019, over half of which promoted conspiracy or fringe content, including racist materials.[44] It is not always clear who develops the original content, since groups circulate much material that individuals in turn do too, often without attribution.

Third, platforms like Facebook automatically tag articles, indicating their popularity through such means as including the number of likes and views. Robotic sites viewing such content inflate the number of likes as well, far beyond the actual number that only humans would have registered. If the proverbial "everyone" likes a story, people are more inclined to look at it, too.[45]

To gain insight into how information changed inside the Internet, I briefly follow one example to suggest patterns of behavior, in this case 9/11. While it is too early to conclude definitively that it is emblematic of other instances of fake facts, conspiracies, and motivations, it appears anecdotally similar to other instances. Recall the event. On September 11, 2001, nearly 3,000 people died in four coordinated attacks by an Islamic extremist group crashing airplanes into New York's two Trade Towers, the Pentagon in Northern Virginia, and a fourth into Pennsylvania. Some 400 firemen, police, and other emergency personnel lost their lives in New York. The United States responded militarily by destroying the government that the Taliban controlled in Afghanistan, which also led to years of warfare there, subsequently in Iraq, and a decade later to the killing of the mastermind of the 9/11 attacks, Osama bin Laden.

In the United States, people were nervous, not knowing if more attacks were coming, while prejudice against Muslims and Arabs increased.[46] This tension spread over the next two decades. No longer was life in the United States seen as safe. Historians know from exploring other periods of tension that urban legends would spread as a coping mechanism for dealing with uncertainties and fears.[47] Legends make it possible to explain a situation, give it meaning, to make sense of it all, regardless of whether factual or not. In fact, people create legends without adequate facts based on their prior contextual understanding of what is plausible. If a company has fired employees on a regular basis for years as its business declined, rumors about new layoffs make sense and become easy targets for narratives that senior managers like to lay off people rather than grow revenues.[48] That happened with 9/11.

Within days, rumors and urban legends (stories, narratives) began spreading over the Internet and continued to circulate for the next two decades. They ebbed and flowed in volume and frequency with which they were read—a great deal within months of the attacks, and less so after bin Laden was killed. They shared common themes, such as *foreknowledge*—"a grateful terrorist" warning American friends who had been good to him to stay out of the Trade Towers on that day; *conspiracy* themes—that Jews knew to stay away from the Trade Towers too, so possibly were involved in the attack, or that the national government's intelligence services knew it was coming but did nothing to prevent these; *pareidol*—the act of identifying something that humans would see as an omen, such as the image of the Devil in the smoke arising from the ruins in New York; *religious themes*—such as the discovery of an undamaged Bible in one of the crashed airplanes; *heroes*—famous actors helping to clear rubble, such as Steve Buscemi (who did not); or *villains*—such as Arabs at a New Jersey coffee shop cheering when they saw the airplanes crash into the Trade Towers (also did not).[49] The events of 9/11 spun off similar false facts and legends in other countries.[50]

There already existed fact-checking websites on the Internet that quickly went to work to separate falsehoods from truth, and none seemed busier than Snopes.com.[51] Between 2001 and 2011, it tracked 176 cases of 9/11 urban legends. Individual queries about specific legends resulted in many hundreds of queries coming to Snopes.com for verification.[52] The largest number of rumors concerned potential new attacks or suspicious behavior. I was interviewed by Fox News and several radio stations in 2002 in connection with a book I co-authored about 9/11, and all the hosts wanted to talk about were these two points, in the case of Fox broadcasting to some 1.7 million people early on a Sunday morning.[53] Since bin Laden had a well-planned, well-funded, and well-executed operation, surely he would be able to pull off additional attacks. Hence the plausible concern expressed by the interviewers and many Americans. Perhaps malls would be next, or big cities, such as the rumored Los Angeles and Chicago. Subways and their tunnels were of concern too. Suspicious behavior proved too tempting to ignore: mysterious rental trucks, crop duster sprayings, hijacking of trucks filled with explosives in Mexico (was it destined for the USA?), Arab women texting in a theater wearing black clothing with their faces covered, and foreigners sneaking across the border into the United States. People waxed about rumors and speculation, feeding their desire for information: letters to the editor, inaccurate newspaper and magazine articles, endless rounds of speeches and lectures, e-mail circulation of fake news, postings on social media websites that continued right through the second decade of the new century, and websites established to nurture the hunger for information.

In going through Snopes.com's postings, William Aspray and I identified other types of information and false facts concerning human interest stories (e.g., heroes and villains), memorials and TV testimonials, celebrations, a vast number of rumors and stories about bin Laden and Afghanistan, inappropriate behaviors, business practices related to 9/11, boneheaded government actions, stories about how Arab terrorists went about their work, how to prevent terrorism, conspiracy theories, misinformation about Islam being a religion promoting terrorism and violence (it does not), and the posting of fake visual images.[54]

Many accurate (truthful) facts, fake facts, and interests represented textbook examples of how conspiracy theories circulated. 9/11 was not new; it was simply a recent one. Lincoln's assassination demonstrated the same behaviors, concerns, and types of rumors; Kennedy's continues to do so. Nothing makes for a better conspiracy than "allegations that powerful people or organizations are plotting together in secret to achieve sinister ends through deception of the public."[55] After 2016, President Trump spoke of his critics within the U.S. Government as part of the "deep state," plotting terrible things against him.[56] Here is not the place to discuss the sociological and psychological causes of these kinds of conspiratorial rumors.[57] But it is the place to point out several features: racial (Jews and Arabs were involved), anti-big government (The U.S. Government knew 9/11 was going to happen, and so too President Roosevelt before the attack on Pearl Harbor by the Japanese on December 7, 1941). It always helped to identify a culprit, then hang on them all manner of sinister plots and actions.[58]

Snopes.com did considerable work to determine the accuracy of myriad stories.[59] Its staff recognized that the Internet could, "be used to spread misinformation as rapidly as accurate information, but all too often the former is set loose to spread far and wide, and correction comes too late, if at all."[60] They addressed a considerable number of stories that were coping mechanisms, investigating rumors that emasculated potential enemies, such as bin Laden's supposed various medical conditions, or that he was dead. People wanted to engage, somehow, in fighting the terrorists, so did by launching rumors. They wanted to carpet bomb Afghanistan, later Iraq, while rumors of General John J. Pershing (1860–1948) killing Moslems in the Philippines before World War I were dredged up as a possible solution to today's problems, making sure that the bullets used to execute the enemy were, of course, coated in pig grease.[61] Because terrorists did not fight for a government in military uniforms, they were portrayed as less honorable and so to be further despised.[62] The Internet facilitated the movement of thousands of rumors across the United States, which now are major documentary bases for those scholars and journalists studying 9/11 and fake facts.

These stories also circulated most frequently at the closest points to the event. So, through both social media and websites, as well as verbally in conversation and local print media, these stories made the rounds in the New York/New Jersey area and in the Washington D.C. metropolitan region. Urban centers became hot spots for such rumors, believed because if New York could be struck, then why not Philadelphia, Chicago, Los Angeles, or San Francisco? It would make perfect sense for bin Laden to do so. After all, if he could make it in New York, he could make it anywhere, parodying a popular song by Frank Sinatra.

Snopes.com and other fact checkers proved successful in uncovering untruths and posting corrections. That their corrections were not always accepted is borne out because falsehoods and rumors kept appearing from new sources and websites in similar and various forms over the years, as had earlier ones about presidential assassinations. But scrutiny of publicly available information had become a staple in information handling on the Internet. Political fact checking, in particular, became a rapidly expanding form of information handling on the Internet by 2010. Aspray and I concluded in a study of the new presence of online fact-checking that it was driven by the reality that American politics had become "increasingly riddled with fake facts" that sprouted "rapidly and widely across the Internet."[63]

The new features of information were the quantity and speed with which these spread. Twitter comments could be created, for example, by President Trump in less than a minute, but then viewed by about 20 percent of all Twitter users, the number who followed his messages, or approximately 14 million people.[64] Aspray identified well over a dozen fact-checking organizations dueling against the onslaught of fake political news, the majority of which came into existence between 2007 and 2018.[65] Nobody seemed immune. For example, from 2018, fact checkers found misinformation coming from Democrats and Republicans, not just from Trump. Caught in the act included Bernie Sanders, Mitch McConnell, and Nancy Pelosi; the first two were senators, the third served as Speaker of the House of Representatives; also by ex-President Bill Clinton.[66] The national elections of 2008 and 2012 stimulated the growth of fact checking, while the elections of 2016 and 2020 ensured it would remain a major new feature of the American information landscape. AI began to be applied to scanning vast quantities of Internet-housed information, using software to scan social media and Google, for example, and increasingly non-political fake facts. Truth seeking was being automated. While fact-checkers, journalists, and academics were the first to recognize the need for checking and pursued it, the public remained outside the fray, largely out of insufficient understanding of how to scrutinize what they were seeing on their screens, or possibly not interested in doing so.

Every time I revised this chapter, there seemed to be fresh cases to consider. For instance, in June 2021, Snopes.com had to investigate whether former President Trump had worn his trousers backward while giving a speech to Republicans at their state convention in North Carolina. Photographs on the Internet suggested he had; Snopes.com investigated and concluded he had not. For two days, every major mainstream media outlet wanted to talk about Trump's pants.[67]

We have the stunning example of events revealed in the Philippines to remind us that problems persisted. Over 95 percent of adults there had access to Facebook in the 2010s, perhaps the most users of Facebook as a percent of the population in any country. The government of President Rodrigo Duterte exercised a heavy hand in running the country, controversially encouraging the killing of drug addicts, and more pertinent to our discussion, aggressively using social media to disseminate false information about its political enemies. One of the president's targets was Rappler, an online news media company in the Philippines, and its CEO, Maria Angelita Ressa. She had been a critic of the government, was arrested and convicted of tax law violations and "cyber libel." The upshot was that despite her legal problems at home, Ressa was awarded the Nobel Peace Prize in 2021 for safeguarding "freedom of expression, which is a precondition for democracy and lasting peace," along with Russian journalist Dmitry A. Muratov for similar reasons.[68] Both had called out corruption and other illegal activities in their nations. Ressa published a book-length memoir describing how her enemies used Facebook and other social media outlets to promulgate mistruths and conspiracy theories, and to destroy the credibility of local news organizations, including her own. She called out Mark Zuckerberg and Facebook for allowing that to happen. Her observation of that experience was that "most of us are predisposed to share what our friends and family tell us," so, "if they receive only fake facts that is what spreads," and in her country that is "exactly what happened."[69] She warned the world that this was occurring in many countries to such an extent that democracies were collapsing, brought down by fake news.

THE PROBLEM OF INFORMATION ILLITERACY AND RESPONSES TO IT

Scrutiny is the core action related to discriminating between factual information and misinformation and other forms of mistruths. This has been an important concern long before the arrival of the Internet. So, one's inability to distinguish between truth and falsehood is a form of information illiteracy. Given the amount of material now available through the Internet that individuals access, online information scrutiny and information illiteracy combined represent new facets of an age-old problem and new ways of shaping information. It is the age-old question, "How do I know that what I am reading is truthful?" "How accurate is Wikipedia?" In the case of Wikipedia, scholars fact-checked this source, comparing it to the rates of accuracy of such venerable publications as the *Encyclopedia Britannica*. The factual error rates of both were roughly the same.[70] Although when a topic is of great controversy, a website containing material about it, such as Wikipedia, can experience many attempts by people promoting bogus data and others to correct these to battle it out, sometimes several times a day at one site. Depending on when a user accesses such a site, they may encounter shifting truths, a result of "revenge editing." Was President Obama born in Kenya or Hawaii? In the 1970s such a question would have been fixed in print, not dynamically changing repeatedly in the course of one or a few days.[71]

Begin with a definition of scrutiny: "the careful and detailed examination of something in order to get information about it."[72] Methods that people use to arrive at truthful, useful information while applied still leave often contested what is considered truthful. If one sees information from the Internet consistent with prior known information or beliefs, they tend to believe it. If the source is a respected one, that helps too, although a fake website made to look like a legitimate one is a constant problem. It is easy to create fake websites that look like, say, that of the *New York Times*, Wikipedia, or a medical association. In 2018, the Pew Research Center, a widely trusted source about behavior on the Internet, conducted a survey that revealed nearly two-thirds of Americans who have heard of media bots thought they were used in some malicious manner. Because so much news comes through the Internet, this presented the problem of having "a negative effect on how the public stays informed."[73] Experts on uses of the Internet had gone on record the prior year as uncomfortable with what was happening. They predicted that more inaccurate materials would continue to appear, that technologies would make it easier for nefarious players to add bad content that could or would not be countered at the scale needed. Labeling content could become a more widely used practice to counter such behavior by websites and social media platforms.[74] All three circumstances occurred in subsequent years.

The old culture of the expert being the one to create and disseminate information was eroding as the dominant provider. Experts were still working but were now joined by non-experts, many times their number, doing the same. Put another way, the decline in their monopoly was destabilizing, with the result that even legitimate truth-seeking experts were now being questioned about their veracity. This became an issue as early as the 1990s with respect to global warming debates in the United States and again during the pandemic of 2020 when the Trump Administration criticized the advice doctors were promulgating out of the CDC. His administration's actions were in response to the experts' contrarian advice about the pandemic that conflicted with the president's political agenda.[75] Gate-keeping editors of newspapers, magazines, and book publishers were outmanned and overrun. Internet-born fact-checking organizations, such as Snopes.com, were too. Even within their own walls at iconic media outlets, problems spilled out. For example, in June 2020, the *New York Times* published an editorial by a conservative American politician that this generally liberal newspaper had not fact-checked, let alone questioned because of its views, raising a row within its own ranks of journalists and reporters. Senior editors quickly had to go public with statements to the effect that their truth-validating procedures had not been followed, at least in this instance. The editor responsible for publishing the piece was quickly dismissed.[76] The political right had a field day with this incident.

Problems with facts concerned at least half the public. People complained that they were getting bad medical information about the pandemic. Americans were concerned, too, about the negative effects inaccurate or misleading information would have on the viability of democracy, a problem dating to the birth of the United States that still worrying the nation. Experts were calling for guardrails on the Internet to protect truth. Facebook refused to participate. Other platforms began experimenting in 2020, such as Twitter (now X), which warned its readers that some of President Trump's tweets had questionable content that violated the site's user policies before letting people read these.[77]

But observers of the Internet viewed these incidents more broadly, and their perspectives aligned with the research Aspray and I have done. Professor and director of the School of Public Affairs at Arizona State University, Karen Mossberger, flagged a wider concern worth quoting at length:

> The spread of fake news is not merely a problem of bots, but part of a larger problem of whether or not people exercise critical thinking and information-literacy skills. Perhaps the surge of fake news in the recent past will serve as a wake-up call to address these aspects of online skills in the media and to address these as fundamental educational competencies in our education system. Online information more generally has an almost limitless diversity of sources, with varied credibility. Technology is driving this issue, but the fix isn't a technical one alone.[78]

From Northwestern University, Mike DeVito accessed the issue from the perspective of the user's skills: "These are not technical problems; they are human problems that technology has simply helped scale, yet we keep attempting purely technological solutions. We can't machine-learn our way out of this disaster, which is actually a perfect storm of poor civics knowledge and poor information literacy."[79] Internet expert Deirdre Williams held emphatically that "Human beings are losing their capability to question and to refuse. Young people are growing into a world where those skills are not being taught."[80] It seemed a given that bad actors would pollute the Internet and that it was up to the rest of society to protect themselves. That is how we arrive at the case for a new information literacy.

THE CASE FOR A NEW INFORMATION LITERACY

A new literacy is needed. As early as 2017, a software learning developer, Julia Koller, explained, "if readers do not change or improve their ability to seek out and identify reliable information sources, the information environment will not improve."[81] She was not alone in her concern.[82] Almost from the beginning of the Web's availability, the new literacy had to include an understanding of how and why websites were and are created and the motivations behind their sponsors, not merely how to make queries and navigate it. We are talking about a literacy different from computer literacy. Computer literacy has long focused on two skills: how to access and use specific technologies and applications, such as a PC, smartphone, or Google, and in a more advanced form, how to program in a software language or to create spreadsheets, for instance. The issue of what constitutes computer literacy remains contested after forty years of discussion but is not the same as information literacy.[83]

The latter combines print and oral traditions, as often people write on the Internet the way they speak, but organize materials the way they might in a book.[84] So Internet gurus want people to understand how it works and who are the players. What do librarians think? After all, they have centuries of experience sorting out the good from the bad. Nicole A. Cooke trains future librarians at the graduate level. She and her students wrestle with the problem of using information in ways contrary to their belief that truth is noble and wiser to embrace than inaccurate or misleading facts.[85] Philosophers, priests, scientists, and educated people around the world agree with this belief. It begins with understanding why people lie and spread falsehoods, discussed in this chapter and elsewhere.[86] Librarians suspect many people cannot differentiate "good" from "bad" information, opinionated statements versus vetted facts.[87] Shorter attention spans also contribute to the problem, since vetting takes time, while quick answers do not, and are more convenient.

It helps to probe deeper to understand the problem, and for that one can be grateful to the Pew Foundation for conducting one of the first studies on this issue. It ran experiments to see how people differentiated between truths and mistruths. The results should leave the reader concerned. Its key finding: "even this basic task presents challenges," because people only identified correctly three out of five statements in each set, "but this result is only a little better than random guesses. Far fewer Americans got all five correct, and roughly a quarter got most or all wrong."[88] Pew's researchers only used political statements. They concluded that for people to discriminate better, they would need to be politically aware, technically savvy, and only rely on trusted sources.[89] One could extrapolate that these three skills would be needed in other disciplines and topics. Literacy, then, begins with organizations, enterprises, and governments; legislators and regulators working to clean up the Internet, pushing out those who would populate it with faulty data. For others, this would be a call to find ways to expand the best practices of the pre-Internet fact checkers, a process that some are already doing on the Internet, notably newspapers and Internet-born fact-checking groups. The challenge is that bad actors come in all forms: government officials, think tanks, lobbyists, politicians, unethical firms and scientists, people ignorant of the truth pushing conspiracies, and their relatives' and friends' unorthodox ideas. Blogs are a particularly large source of misinformation that needs to be mistrusted because they are such an easy entrance ramp onto the Information Highway. Bloggers pick and choose existing content, repackage it, and send it on its merry way to pollute the thinking of other people, sometimes millions of them.

Literacy can be assisted by the work done by reputable news outlets fighting an epidemic of fake news. Earlier, we noted how many times President Trump spoke mistruths. That was possible to state because newspapers like the *Washington Post, L.A. Times,* and the *New York Times* fact-checked his comments, providing immediate corrections and literally keeping score. Those kinds of activities are cropping up all over the Internet around the world.[90] To be Internet literate, people need to pay attention to those initiatives and learn from them. Consider it a form of information hygiene. It needs to be done regularly, like brushing one's teeth.

A third issue involves the fact that there are adults and younger people who have not learned to differentiate one from the other. They may not have been taught this at home, school, or college, no matter as such individuals are known to all of us. Ultimately, this goes to the core of what formal education is about: the harvesting of truth and discarding of falsehoods. It has to be the individual, not some institution, who is responsible for their own intellectual integrity. It can be trained into an individual, just as handed down values are in families, religious values to seminarians and congregations, leadership and integrity to military cadets, and corporate cultural practices to business employees. I use the phrase "trained into" to suggest it has to be a proactive, well-thought-out, deliberate action, not something that happens as a byproduct of other pedagogical activities. Bad information comes piecemeal into someone's life, just like Maria Ressa pointed out happened in the Philippines. Accurate and inaccurate information spreads quickly, regardless of whether it is intentional or not. So, this is a serious problem for individuals and entire societies that must be acknowledged as occurring and be addressed. There is more than truth influencing one's behavior, with profound effects on societies at large. But, to be even handed about the matter, how technology developed also contributed to the variability of truth, such as the use of algorithms and modern society's adulation with quantified information.[91]

The reason for the importance of one's personal responsibility is that in many societies, individuals are held accountable for managing their own behavior, as in the United States. If that is the value individuals and the nation embrace, then it stands to reason that responsibility also includes learning how to differentiate truth from falsehoods. The American political scene, in particular, requires personal responsibility exercised through the ballot box, which since the late 1700s has required people to be informed about issues of the day and the candidates standing for election. While that responsibility has recently devolved into more tribal behavior than the Founding Fathers would have imagined, the fundamental requirement to personally take charge of being informed to vote wisely did not go away.

That responsibility applies close to a second one: the role of commerce, since residents in the United States think they live in a land of economic opportunity. Legal and regulatory reforms to keep access open to information continue to exist. The public sector's explorations of behavior by Facebook, Google, and others are emblematic of that long heritage continuing in our time of the Internet.

Finally, critical thinking skills are needed, skills that transcend every prior and current information platform. For nearly a century, IBM had a bumper sticker corporate motto: THINK. When originally conceived, IBM's leader, Thomas J. Watson (1872–1956), had it in mind that employees should think in a certain, "IBM Way," but also that they should cogitate over their customer's problems, working out what the critical issues and opportunities were that his machines and computers could resolve. Within a generation of people living under THINK, thought.[92] By the time I joined IBM in 1974, I quickly realized IBM was an organization filled with as many intellectually thoughtful people as any university, just as well educated, and capable of acting on their discernment. I was expected to do the same. The elements of critical thinking involved discriminating about sources, language, argumentation, and content. All are present on the Internet, some packaged as in prior times (e.g., verbally, textually, PDF files) and some in new ways (e.g., as in the style of badly written Letters to the Editor known as blogs and sound bites on Facebook). Foundations, scholars, and teachers have tumbled onto the value of this observation and started to present prior good critical thinking skills within the language and context of the Internet.[93] That needs to continue.

As with the case of teachers, librarians remain some of the most underappreciated information workers of our time, but they quietly continue to address core information issues. They too want to continue exposing fake facts and encouraging better critical thinking habits. To that end, some have been working on what they use among themselves as a new term, *metaliteracy.* Its purpose is to merge information technologies with other types of literacies. Both purveyors of bad information and those scrutinizing for accurate information can, of course, use the concept. A librarian explains:

> Metaliteracy asks us to understand the format type and delivery mode of information; evaluate dynamic content critically; evaluate user feedback of information; produce original content in multiple media formats; create a context for user-generated information; understand personal privacy, information

ethics, and intellectual property issues; and share information in participatory environments.[94]

That would appear to be a mouthful to say, less so for a corporation, government agency, librarian, or computer scientist to implement. But take a few minutes to consider each part of the sentence, and it should become clearer that this new literacy is actually an old one in seven parts. As in a century or two ago, it calls for each to be applied simultaneously. They do force one to think of information literacy more broadly than simply knowing how to Google information. It is a call to judge what they are reading, where it came from, and the motives of those who created the content.[95] It can all begin when a child begs for more "screen time," when a teacher issues a tablet to a 10-year-old student with which to learn about animals and fauna, when an employer gives a young employee access to the company's online files, or when a university online access to many hundreds of databases to its students, faculty, and staff.

It is now time to pull together a constellation of insights about how people deal with information: historical patterns of use, how our brains process facts, putting into perspective information ecosystems, the role of computers, and settling on a context in which to deal with today's informational realities. Those are our challenges for the rest of this book.

Chapter 6

A Way to Look at Information Today

> *. . . whole living organisms and their environments will be computationally modeled, with all molecular categories taken into account simultaneously.*
>
> —Team of biologists, 1992[1]

Information as a topic, issue, or concern originated in some of its contemporary discussions in religious circles centuries ago, spread to secular scholars and professions before the Renaissance (e.g., law and public administration), and afterward to artists and proto-scientists. By the 1700s, expanded bodies of organized information were increasing in variety and quantity, and people began segregating these into disciplines. The latter then fragmented into more specialized ones, such as subfields of science, while the industrialized world became more reliant on scientific and engineering information. After the wide appropriation of electricity, followed more significantly by computing, humans' dependence on organized bodies of information increased sharply. That trend exposed conversations about the nature and use of information held within academic and professional disciplines and within specialized and widely available information infrastructures and publications. Today, the general public participates more openly in discussions about the role of information in society, the economy, their jobs, and personal life, but largely through social media and political activities, not necessarily just as scholars or as part of their professions.

Information's visibility did much to change interest in the subject. As noted in earlier chapters, PCs and the Internet made vast quantities available in diverse forms and content essentially to anyone with access to such devices, including smartphones. No longer did experts control the quality, amount, and types of information available to anyone. "Anyone" could add information (accurate, false, or irrelevant) to the fountain of human knowledge. So, in addition to experts specializing in specific bodies of information from the eighteenth century busily creating and changing it, everyone else seemed to be doing the same. As a consequence, how one respected and used information changed, depending on where they perched within the enveloping information ecosystem. The amount of time and money spent in nurturing and relying on information, as measured by the amount of "screen time" consuming their days, for example, or what they spent on their information infrastructures, quantified their increasing dependence, especially since the 1980s. In the past two decades, computers and their software have joined in the fray: artificial intelligence (AI) creating "new" information and making it so easy for anyone to access.[2]

Since at least the seventeenth century in Europe and in North America, farmers, day laborers, and anyone with a modicum of reading skills increasingly became dependent on bodies of information created by experts in order to go about their work and private lives. By the time of the Second Industrial Revolution in the United States, say, in the 1880s, a farmer in Wisconsin might be visited by an employee of the University of Wisconsin to educate him on the latest scientific findings about how best to raise corn or treat a farm animal's illness. His wife would probably have been consulting the *Farmer's Almanac* or other published sources regarding weather forecasts, as most likely would have her mother and grandmother if raised in the United States. Today, a farmer might talk to his smartphone to ask for instructions on how to repair a piece of farm machinery, while a drone surveys his fields to determine what needs watering that day. The same behavior existed in factories, in all blue-collar trades, and in the arts. For people in all walks of life and ages, information had become practical tools made relevant to whatever their needs were by the specificity of the facts and advice provided, and by its convenience. Even children as young as seven were accessing the Internet, some for schoolwork (at least during pandemic remote classes).

Those are broad historical trends that bring us to the third decade of the twenty-first century. Historians have documented these trends in the life of information to such an extent that we have to ask: How should we look at information in our time? It is a question relevant for any academic, information-intensive profession, and for the public at large. As previous chapters demonstrated, the conversation, issues, and breadth of the topic are extraordinary and still evolving. The diversity of "content" is difficult to comprehend, while a combination of Trump Era relations with truth and time becoming increasingly available during the COVID-19 lockdowns of 2020–2022, forced entire societies to engage with information in far more public ways.

The consequences of such altered behavior are not fully obvious, although consistent with information creation, opining, and speculation.

Ancient Greek oracles predicting the outcomes of battles could be forgiven if they concluded, as did a philosopher at Princeton University, that there was much BS.[3] But entire societies—and you, the reader—are immersed in the debates. This introduction to our almost final chapter is a long way of saying one might need some process for how to deal with it, or at least the issue to keep in front of mind as one engages with related topics. Before one jumps to the conclusion, *aha!*, that this chapter is about to present yet another theory of information or paradigm, I already partially snuck that in within earlier chapters and prior publications. The best that can be done here is to point out that there are new influences in what constitutes information and how one might engage with these.[4] The epigraph introducing this chapter is more than a hint of the changes we face: biologists weighing in on the definition of information, and the impact of computing in the development and use of that going forward. At the moment, however, biologists may have served up a positive dish of hubris. However, they, psychologists, and neuroscientists are doing more to help us understand how humans deal with information than experts in other disciplines.

Branches of sciences never coordinated and used universally accepted techniques; they had too many diverse issues and data requirements to contend with. As a result, fashions in methods and thinking evolved. Chemists rode high in the late 1800s; today it is the biologists, or at least those from certain branches of their discipline. Entwined in their worlds were individuals and groups that made it necessary to discuss Big Data in chapter 4 and false facts in chapter 5. The information ecosystem had evolved into a complex one. The epigraph demonstrates that the subject had become a team activity, one requiring the participation of multiple researchers to flush out insights.

First, we need to settle on a working understanding of how human brains process information, if ever so briefly. Then, we should view information and its role through a series of perspectives: information ecosystem, a mixture of information technology's methods of thinking combined with notions of ecosystems, because we will need a way to think more closely about the role of AI and robotics in our last chapter, and wrap up with a broader way of understanding our politics, beliefs, and secular realities—the role of context—as we work to understand subsequent evolutions in our understanding of information. For if there is one certainty one can extract from prior historical experience, it is that we are encountering new features of information, which, while anticipated as early as a century ago, are just becoming realities. I may be leading you through an epistemic view of information or simply using the analogy of an economic system to affirm the value of thinking in terms of ecosystems and networks populated with humans using their brains in ways we did not understand a generation ago. All of this is happening within the context of an enveloping global ecosystem comprising far more elements revealed over the past century. If so, that is the essence of my theory of the role of information as best I can offer so far. To accomplish our tasks here, we must simultaneously keep much in mind.

HOW BRAINS PROCESS INFORMATION AND SHAPE WHAT ARE FACTS: THE POWER OF BIAS

In the past half century, multinational armies of neuroscientists, biologists, psychologists, psychiatrists, philosophers, historians, economists, and sociologists have cumulatively enhanced our understanding of how humans accept, process, and use information.[5] The older one becomes, the more their brains change, such as by forgetting more compared to when their memory was sharper years earlier, later thinking slower and relying more on experience and "gut instinct" rather than documentation, or by relying on preexisting beliefs against which to assess the validity of contradicting information. Later, I return to the issue of context as an important influence on how people deal with information; it is enough here to observe that older individuals rely extensively on that feature of their thinking. Our understanding of brain functions has reached a point where discussions about how humans deal with information should take into account such findings. It all begins with the brain, followed by any intellectual constructs one can encounter in the rest of this chapter, if the discussion is to remain rooted in the anthropomorphic.

Pay more attention to what neuroscientists and psychologists are learning, regardless of our individual expertise or profession. While the volume of studies about the human brain is large, one useful trend to applaud is that many scientific findings are making it into the larger space of society, thanks to recent publications about the topic aimed at the public and across disciplines.[6] These provide sufficient information about brain functions for the public to use, the idea of "good enough," to go about their activities. Students of information will increasingly need to consult such publications to benefit from research findings. Paying attention to the growing scientific understanding about the human brain represents a partial break from twentieth-century behavior when few research findings about it were widely accessible outside a few academic disciplines.

So why do students, workers, parents, and management need to add this kind of information to their intellectual toolboxes? Because how one processes information influences what facts are created and used, and what conclusions people draw from these. Their responses shape how information changes. It is also useful to understand the effects of bias on one's views because we process and apply information through the lens of our biases, even more so than through "scientific spectacles." One conclusion from historical experience is that people will continue to develop more and varied bodies of information that in turn will affect how they function in their individual information ecosystems. Even further into the future, AI software will do the same. The physical center of that human activity is largely in the brain, although other parts of the body, too, must respond to the realities they face per instructions from the mind. The accumulation of brains affects how information ecosystems function, although as with any ecosystem, other realities impinge too, such as changing weather and climate, activities of other animals, and so forth. In other words, context remains important.[7]

Is this discussion the same as one might entertain regarding "situational awareness"? This phrase captures the idea of someone (or a group) aware of an immediate set of circumstances surrounding them that take into account time and space. Examples include knowing what automobile drivers near you are doing as you drive, including taking into account such other circumstances as road surface or weather, or an airplane pilot doing the same regarding 360 degrees around her aircraft. Such awareness can lead to immediate decisions in defense of life (such as what a law enforcement officer may be confronted with or a soldier on the battlefield), avoiding danger (such as crashing a car), or dealing with a medical crisis. Situational awareness calls for immediate decisions and taking quick actions in response to a reality in crisis, followed by subsequent less time-sensitive ones. The concept originated in the U.S. Air Force for combat operations, but academics began studying it in the 1990s.[8] It is about understanding what is going on to shape a response. The brain must operate as fast as it can, digesting small amounts of information and reacting quickly. There is no theory building or scientific analysis going on with situational awareness. Context may play a role, such as a soldier understanding he is at war with those from an enemy state, but he has to operate more urgently than contemplating context requires. His context was integrated into his thinking long before the immediate realities he is facing. So the context is different and supportive.

What can briefly be summarized here? First, identifying with a community's values and bodies of information has long been recognized as profoundly influential on our thinking. Scientists believe in the use of scientific methods for uncovering information, citizens value patriotism in support of facts reinforcing a nation's way of life, or that our *alma mater* has a superior football team. The same applies to such influences as ethics, religious beliefs, and political affiliations.[9] Sound reasoning and communication reinforce the collection and use of a community's body of knowledge.[10] However, each of these can be faulty, particularly if one is passionate about an issue, in which case their reasoning usually does not help in sustaining facts but only to twist them.[11] But one of the most influential functions of the brain affecting how one defines, collects, and uses information is bias, and that aspect of thinking has attracted an enormous amount of research since the end of World War II.[12]

Bias is an inclination favoring or dismissing an idea or group of people: that liberals are more open to new ideas than politically conservative citizens, that Wisconsinites think automobile drivers from Illinois are terrible, and that North African immigrants in Europe are probably pickpockets. The kindest observation one can make of these is that such images are, at best, only marginally—not generally—true. They can be negative, but also positive: that Germans are well organized, that women are more nurturing than men, and that Hispanics are a friendly lot. Often stereotypes are called prejudices; psychologists see them as biases. We are taught some of these in childhood, such as to favor people "like us" more than those who are "different." Others include that economic historians have more to teach us about how to make money than a political science professor, or university students are inclined to be offended at the thought of listening to conservative speakers. The list of examples can be endless. These are biases that subconsciously or overtly affect how one views evidence, facts, or circumstances. They certainly affect how people apply context in their own thinking.

Students of human thinking have identified some one hundred of these.[13] A bias does not require commitment to truth, evidence, or to be empirically grounded in reality. Everyone is capable of believing any nonsense or untruth they want and of taking actions based on these beliefs. We can be overly optimistic about succeeding in our ventures, or constantly negative. Availability bias might involve someone thinking that child kidnapping is always a likely possibility. A familiarity bias used by politicians and propagandists is the act of repeating something enough times that it becomes believed. Vaccines reported to cause genetic diseases, if repeated enough times, cause some parents to resist having their children vaccinated against COVID-19. People believed that the sun rotated around the Earth for centuries because it was always declared as truth, even when it was not. There is the behavior of someone believing they understand better how you think than you do (known as asymmetric insight), or believing they have superior information compared to what someone else has. A frequently seen bias is when someone overemphasizes a piece of information when it is the first fact offered, instead of waiting to see all the evidence before deciding which one to emphasize. I might be doing that here by suggesting biases affect our treatment of information when other options are available.

Framing can be a bias—one that I admit to exercising when suggesting looking at information through the metaphor of an ecosystem. Competency bias is extremely serious because we live in a time when professional and academic credentials are so highly respected. I have a PhD in history, so I must know more than someone who does not about, say, what really happened during the Vietnam War. Yet, in truth, I have not conducted empirical historical research on that war. I just read the same books people who lived through the period did. We have all come across individuals who promote their expertise as evidence that the facts they use are *the* accurate ones. They could be wrong, as it was eventually discovered that Sigmund Freud did not have it all right, or that Albert Einstein may have missed a few points too.

An elegantly named one—perseverance bias—is normally seen as someone just being stubborn about not changing their mind on an issue. Show a Trump supporter a notarized copy of President Obama's birth certificate from Hawaii, and if they still believe he was born in Kenya, you have an example. I am guilty of this kind of bias, too, because I still believe that people are fundamentally good, but the older I become, the more bad people I see running around.[14] One sees this bias when a friend or relative refuses to "listen to reason" or to "face the facts," or is in "denial" about some circumstance. It is even worse because, as one observer noted, our biases "blind us to our biases."[15] Our individual collection of biases makes it difficult to think clearly and, just as bad, not to realize that we are doing so.[16]

The case for bias stems from early humankind's need to respond quickly to dangerous environmental threats, such as a large animal wanting to eat them. Quick response requires less thinking and more reaction, ergo the use of biases learned (hopefully) as useful defenses. Soldiers are taught, for example, that the enemy will want to kill them; therefore, when they see one on the battlefield, their impulse is to quickly kill them. Yet, the enemy soldier might be a medic just trying to help the wounded from either side. Scientists speak of these intellectual shortcuts as heuristics: making decisions by relying on instincts, intuitions, rules of thumb, or gut feelings. Your sense that liberal politicians are more inclined to raise taxes and spend money may sway you even though you do not know for certain that a liberal politician standing for election is a spendthrift. All of this is a problem when, as Brookings Institution's Jonathan Rauch observed, "in a complex environment" when we are called upon to "make sophisticated judgments, our biases often lead us astray, while desensitizing us to the possibility that this might be wrong."[17] As the amount of information created and used increased over the centuries, this problem lurked in the shadows of our cognitive activities.[18] Historians and social scientists have yet to arrive at a thorough understanding of the effects of such behavior on all manner of information.[19]

How one collects and uses information is affected by biases. Believing in a group's points of view makes possible membership—anthropologists speak in terms of tribal identity—which for well over a million years provided greater security than living alone. Religious beliefs fold in here as an example; so too identity with a particular political party.[20] Hearing from only a single group the information one receives about an issue is the echo chamber phenomenon. In the United States in the 2010s, receiving all one's political and economic news from either Fox News (conservative) or MSNBC or the *New York Times* (liberal) illustrates this behavior in which an individual keeps hearing the same facts, reinforcing beliefs and information they already trust.[21] The public reflects such behaviors through their use of social media. Academics behave badly too when some scholar turns his debate into personal attacks, as opposed to maintaining a discussion about an issue through the winnowing and sifting of evidence. Cancel culture is the current moniker for such behavior, and these days it seems most widespread among university students deploring giving an unpopular speaker the opportunity to explain his or her point of view, or parents shouting down a school board.[22]

It seems everyone enjoys a good conspiracy theory, which incorporates many biases, fake facts, and gut feelings, all rolled up into the vicarious experience of a plausible explanation for something. A half century after President John F. Kennedy's assassination, people can't seem to get enough of whether it was one or multiple shooters who did it.[23] Toss in a rumor or two, and one has the basis of a good conspiracy. My favorite: a company with a history of laying off employees when someone speculates that their office will be targeted next for similar treatment. Today, American conservatives think some "deep state" inside the national government is trying to do evil things, while Hillary Clinton still seems to find time to run a child-kidnapping ring out of a pizza restaurant.

Yet, the creation of much information was the result of good observation, rational thinking, and familiarity with other reliable evidence—in other words, the application of objectivity. Scientific methods of the past three to four centuries made that possible, despite the continuing effects of biases. But beyond the role of scientific ways of thinking, people use information to deal with immediate circumstances. As two psychologists explain, "People are primarily designed for action, not for listening to lectures, not for manipulating symbols, and not for memorizing facts."[24] That is why the collective knowledge of a community is often better than what one individual knows (too much out there to keep track of by a person) and why we need digital files and books too.

That collective knowledge plays out in myriad ways. However, at the mundane level of, say, work practices, it is clear. Humans in almost every profession have had to increase their consumption of information over the past 150 or more years. That is how one creates specialists in ever-narrower fields of information. Getting any work done more complex than one's knowledge supports has led to the ever-increasing use of collaboration. That is how it was possible for humans to design, build, and operate commercial aircraft, send people into space, or build global supply chains for multinational corporations. When teams are created with each individual as a master of a unique body of information, the quality of the group's thinking and use of information can be enhanced.

AS AN INFORMATION ECOSYSTEM

There are many ways to look at how information increased and changed, as well as how people appropriated it over the past two centuries. Several are relevant today. An emerging concept with various names combines information ecosystems and infrastructures, both often called networks, depending on which discipline's vocabulary one favors. But they are marching toward a common worldview: that information is dynamic and somehow interconnected with multiple bodies of facts and electronic means of making these increasingly available to people, software, sensors, computers, and industrial equipment, which two communications experts neatly called "an environment" resulting from "the rise of the digital in the contemporary world."[25] Various approaches are increasingly interdisciplinary, while methodological problems, nonetheless, make it possible to further the study of social spheres. In turn, social spheres are increasingly an aspect of the study of information. As students of information's transformation plow further into the topic, their perspectives are changing. We do not review that work here, as I and others have done that elsewhere.[26] It is the suggestion that information, its features, and role continue to evolve, seemingly toward some new paradigm, or to use a philosopher's phrase, toward being on the "cusp of consolidation," which, however, it never quite achieves since it continues to change.[27] My own views have evolved too, an experience that informs me about the nature of information and its continuing transition. This is why this book began by announcing that I would try to remain focused on anthropomorphic views of information, although our brief discussion later about AI and robotics in the next chapter represents a potential departure from that commitment. I could not have imagined making a statement about limiting my work to human considerations just two decades ago.

Picking up on the discussion about information ecosystems initiated in chapter 1 and periodically in subsequent ones, the argument that information is situated in some ecosystem originated in the nineteenth century. It was reinforced by the emergence of clusters of related facts, such as those about medicine, economics, or business, and then was buttressed by the emergence of professions, discipline-specific associations of experts, and a variety of journals and other publications focused on specific bodies of knowledge. These various developments became evident by World War I. Then, over the next half century, specialists built up and relied upon increasing mounds of information on the back of those earlier trends. Recall that after computers made their appearance in the mid-century, the accumulation of even more specialized bodies of information and their users sped up, augmented by the ability to share bodies of data across distances by the 1960s through telecommunications, and soon after with desktop computers (1980s), then portable ones. But the real breakthrough that led people to think about information ecosystems came with the wide diffusion of the Internet. This new behavior became evident among researchers, professionals, institutions, and the public at large. The century-wide awareness and growing interest in the nature and use of information unfolded at different times and in different ways: to telephone company engineers in the 1920s-1940s, biologists beginning in the 1950s, economists in the 1950–1960s, social scientists in the 1970s, and historians by the turn of the century. The research done on the history of information presented in my *All the Facts* and in *Birth of Modern Facts* documented these trends.[28]

Why did the Internet trigger new thinking about information being more than collections of facts stacked up in piles of publications or now ensconced in computers? There is much speculation about the answer to that question, but people sharing information as a personal act of engagement parallels what happened with the establishment of post offices and public libraries in the nineteenth century in the United States, for example. It became obvious that such institutions worked because they were interconnected with each other, most obviously the post office, and later national telephone services. Scholars saw that too, as the use of the Internet diffused rapidly by the early 2000s. They began commenting on the interconnectedness of networks, people, bodies of information, and the emergence of an information cocoon blanketing the world. Initial discussions of networks and ecosystems evolved into more about frameworks, of intellectual scaffolding (theories) all intended to explain how people corralled together bodies of information and shared them around. As scholars became more familiar with the concept of the Internet, they began to think of information ecosystems—still not always called that yet—as a metaphor for how society was gathering and using information, now too called data, and still knowledge. These used such pre-Internet era vocabulary, each with its own implications and contexts.[29] Much was loosely conflated and intermixed, such as data and facts, or information with knowledge.

Most of the conceptual thinking came from the humanities, social sciences, journalism, and business schools. However, almost invisibly from the 1950s to the early 2000s, biologists, too, confronted information constructs while studying feedback behaviors of living matter and DNA, of course. Even scientists not in biology began noticing what they called "systems," most obviously how weather worked.[30] But biologists concurrently with computer scientists specializing in telecommunications concluded there existed *de facto* information ecosystems visible once tools needed to visualize them became available, such as more powerful microscopes and modeling software.[31] More significant, they began concluding that living matter was, indeed, information. Telephone engineers had reached a similar conclusion about electrons decades earlier.

What happened? The notion of information ecosystems had moved from being convenient frameworks and metaphors to realities to such an extent that even experts on non-living things began to identify information diffusion among inanimate objects. Geologists discussed the movement of rocks, tectonic plates, and communications to parts of the earth via temperature changes, now expressed as "signals" (i.e., communications, feedback).[32] It seemed physical matter was information and agents moving it about. The examples that caught the public's eye in the new century were studies about how trees "talked" to each other via roots, conceptualizing trees and other plants as ecosystems with non-human-like communications sharing capabilities.[33] Taken as a whole—plants, animals, people, and Earth—could now be seen as interconnected ecosystems laden with, or defined by, information. They were described as dynamic, dependent, and affected by other participants. People increasingly saw such activities as measurable, hence understandable when viewed through their "scientific spectacles." In short, humans no longer monopolized and operated all information; everything else apparently did too. That shift in worldviews may eventually turn out to be as profound a change in perspectives as those brought about by Newtonian scientific worldviews several centuries ago, or simply an overreach or a way station to a more realistic perspective. Time will tell.

But not everything about information ecosystems—ecologies—was identified in the 2000s. The process of identification's evolution perhaps snuck up on humans over time, but once started, it became obvious. Much of the conversation, hence identification, of ecosystems was initially rooted in discussions regarding computers and telecommunications. Two scholars in the 1990s demonstrated that behavior with their definition of an information ecology: "a system of people, practices, values, and technologies in a particular local environment." Their characterization, while conventional, also sported a narrow view: "In information ecologies, the spotlight is not on technology, but on human activities that are served by technology."[34] Today, more machines and sensors are active on the Internet than humans. Increasingly, software determines their uses of data and communicating activities without human intervention. Limiting discussions to the anthropomorphic admittedly narrows the scope of one's understanding of today's information ecosystems.

Nonetheless, descriptions of information ecosystems from the 1990s have held up for three decades. Common features include the realization that interrelationships exist among all components in an ecosystem. Today, one would identify more of these than, say, at the end of the 1990s, as more participants are recognized. Each player is active, along with their information, as they move about affecting activities and shaping the nature of others' information. Put another way, information is not static; it remains constantly active. Each type of information occupies a niche in an ecosystem, just as worms in the soil or birds in the upper reaches of trees. Data from sensors might never be anthropomorphic and always electrical impulses, rather than databases a human can read. In a healthy information ecosystem, one would expect to find different types of information and habitats (e.g., books or smartphone files), while users operate in both a complementary way and some understood harmony. An example is what happens in a library when librarians handle rare books, while the rest of us usually read modern editions sitting on shelves.

What is an ecosystem without diversity of information or inhabitants? Cohabitation is a major activity in the creation and nurturing of an ecosystem. Monkeys need banana trees, lions need other mammals to eat, and humans need information in the form of books, journals, and online files. Students of information ecosystems find that users and habitats of information evolve over time and so do their information. In 1995, one had to go to the library to find birth and death dates of semi-known luminaries; today, Wikipedia provides that data, with the result that historians are more likely to include such facts the first time they introduce an individual in their text than before the Internet. Trivial, you say? Not really, because now we are more sensitive to how old a historical figure was when they did what interested an historian. There is a difference, therefore, in how we approach the accomplishments of Mark Zuckerberg (b. 1984) when he established Facebook, barely in his twenties; compared with Jeff Bezos (b. 1964), who was in his 30s when he created Amazon; or Joe Biden (b. 1942), who became president of the United States in his late 70s. Finally, keystone technologies and containers affect what kinds of information exist in an ecosystem: books, PCs, Internet, DNA, databases, magazines, computers, or library buildings. The list is long but tends to have dominant forms, such as DNA in living cells or data in digital databases. Ecologies or ecosystems imply coordination, constant activity, and changes, each affecting the use and shape of information. Biologists suggest that information is constant activity, while electrical engineers and computer scientists have been arguing that case for decades.[35]

Once we recognize that information ecosystems exist and are home to active agents, all manner of insights emerge. Far more than the phenomenon where everyone seems to be driving the same model as my new car, more important and permanent observations become evident. For example, it is now nearly impossible to think of information ecosystems as two-dimensional static descriptions of information, even though such explanations make it easy to fill books with clever figures attempting to simplify complex concepts.[36] These must now be seen as three-dimensional constructs, because that is what they are, which should make proposals about the presentation of space, data, and truth more aligned with the arguments proposed by Edward Tufte, or geneticists.[37] Psychologist Jonathan Haidt brings us back to the anthropomorphic role in an ecosystem: "each individual [is] as being limited, like a neuron," which, "by itself isn't very smart. But if you put neurons together in the right way you get a brain; you get an emergent system that is much smarter and more flexible than a single neuron."[38] Put more people and other active vessels of information together and one gets an even more effective (think smarter) living entity. This behavior is the ultimate rationale both for having an information ecosystem and for why one should understand the role of facts through such a prism. Insert into this model—reality—of the world's scientific methods of investigation and values of truth, and one moves forward in understanding how information and its users function. Will we need holographic books someday in order to explain what is happening?

Most centrally, humans increasingly saw themselves enveloped by information as early as around 1900, and as a twenty-first-century observer pointed out, "that we find ourselves enrolled in a thousand databases. Who are we without all these identifiers, numbers, and other bits stored away in countless many data warehouses?"[39] In short, we individually, too, became information ecosystems and had been so since the early 1900s.

One reality-based piece of information affects others, generating further action and consequences.[40] It then remains to accept objective-seeking information-hunting practices to thrive in an information ecosystem. Philosopher Helen E. Longino offered an explanation for why: "the greater the number of different points of view included in a given community [think ecology/environment], the more likely it is that scientific practice will be objective."[41] Her idea suggests there exist conflicts, rivalries, and competition in the information ecosystem as in other realms of reality. Allegiance to a perspective about reality is strong, as is groupthink. These behaviors either protect a species through collectively learned lessons of survival or lead to disaster, such as happens to cult groups and to those who deny changing realities (e.g., effects of global warming on Earth's climate). The point is, to be slightly Zen like, what one might see could be interesting, but it is what is not seen that probably makes it useful. It is an observation proffered by Bonnie A. Nardi and Vicki L. O'Day when they explained that a bowl can be attractive, but it is the empty space within it that makes it useful for, say, storing food.[42] Theirs' is an important observation because information ecosystems are normally not visible, such as accumulating Big Data, due to electronic sensors buried underground, bolted to satellites, traveling as electrons, or nestled in neurons. Pause for a moment and think, has anyone ever seen software functioning in a computer? Unless one could watch the electronic dots and dashes representing software instructions pass by on some oscilloscope or other measuring device, the answer is no. Yet software exists and affects much of what we do most profoundly.

As one moves from metaphor to the realization that information ecosystems are as real as living matter, and that their activity is all about moving information to where it is needed to sustain life, systems thinking, discussed in *Birth of Modern Facts* and again briefly revisited in this book, provides a holistic view—think of these as tools and methods—and is also increasingly seen as useful. We will not, here, engage in a debate about whether that is the same as or different from scientific methods and values, which experience suggests are complementary. I simply point out that they reflect a way of studying features of information in most academic disciplines. It is a further refinement of how information is created, changed, and used that goes beyond Newtonian models of the machine and more toward the analogies served up by biology.

AS A MIXTURE OF INFORMATION TECHNOLOGY METHODS OF THINKING COMBINED WITH NOTIONS OF ECOSYSTEMS

As computer (information) technology (IT) diffused throughout industrialized societies, developers and users of these new machines and software formulated ways of thinking and talking about them. Their vocabulary increased by thousands of words.[43] They also borrowed from telegraphy and telephony concepts about data sets, networks, and other notions about groupings of machines and software in the 1940s and 1950s, of "systems" meaning communities of hardware and software in the 1960s and 1970s, reaching back, too, into the nineteenth century for intellectual practices and models. By the end of the 1970s, and most dramatically during the 1980s, their language seeped into the wider world of scholars in various disciplines, and into the general public's vernacular by the 1990s. Diffusion of the Internet turbocharged these behaviors. Today, one can *Google* (i.e., look up something) or *text* someone. Earlier, in the 1980s, managers began speaking and thinking about *processes*, rather than workflows or tasks, as had been the norm since the early twentieth century. Because so many components of a computer's operations had to be *compatible* with each other and able to *collaborate* or be *integrated* to work, thinking about ecosystems using such words became a useful mindset by the 1960s that endures today. Notions, with their IT language, penetrated into all manner of discussions about social ecosystems and discussions about the history, form, function, and consequences of information. Scientists are the most extensive users of such concepts, with social scientists, media experts, and business professionals and managers right behind them.[44]

Such notions also reveal consequences. When one development or concept is created, it can affect something else. French sociologist Jacques Ellul made a similar point about the evolution of technology in which one development for a specific purpose ends up resulting in some unanticipated different one, such as how we understand a situation or how one approaches a task. He explained with an example: "machines that use perforated cards affect statistics and the organization of certain business enterprises. Conversely, some kind of social technique (for instance, full employment) may entail an improvement in the techniques of economic production."[45] It is why he and others could argue that technology and information did not increase just linearly but geometrically, in his words, "without deliberate will, by a simple combination of new data, incessant discoveries take place everywhere; and whole fields are opened up to technique [his word for processes] because of the meeting of several currents."[46] All may occur unevenly, which may go far to explain why there are various consequences.

There is one idea that has served computer scientists, IT builders, and their users for over seven decades. This idea is easy to describe, can accommodate much complexity, and is useful as we discuss information in the remaining parts of this chapter. Essentially, it is a *system* that works, not just the individual components of it, and all parts within a system must operate in a coordinated manner. Just as the human body consists of a system of bones, organs, muscles, and so forth, all coordinated by a combination of the brain and the nervous system, and increasingly, as now understood by biologists, by information embedded in neurons and all living parts of any living creature, so too do computers. Thus, for example, a computer consists of data input equipment (e.g., a terminal or file on a disk), a processor that "crunches the numbers," and output equipment that receives the results of such processing, such as another disk drive or a PC screen. That system requires operating software to coordinate all these activities, such as Microsoft's Windows or, on a large mainframe, IBM's operating system software.

Big or small, there are also utility software programs that make it possible to translate programs into terms the machine understands, others that organize and present data in ways comprehensible to humans and machines, and networking software that exchanges information with other systems. Perched on top of that software may be other systems that connect smaller or distributed systems comprising hardware and software, such as PCs, laptops, and smartphones. Everything has to be able to *talk* to each other to function as desired. In the mid-1940s, this was such a poorly understood idea that one of its earliest modern proponents, John von Neumann (1903–1957), sketched it out in relatively simple language. That act contributed to scientists, engineers, and historians reaffirming his brilliance; today, most users of computational devices accept such an explanation as obvious.[47] That ability to think of a model or piece of information as obvious also reflects the evolution of our thinking that came about as a way to deal with today's information. Von Neumann's handiwork is an example of someone able to think conceptually, that is to say, scientifically.

Almost from the computer's earliest conception, figures (images) presented the concept in terms relevant for each decade, such as for hardware and software of the 1960s or 1980s, and so forth. These were mechanisms for illustrating their interdependencies and where they connected with each other. That method is now used increasingly to explain information ecosystems. Look at figure 6.1. Read it from bottom to top, with the most fundamental basic technologies lowest, in this case, the computer machine. Next up is the software that inside the machine makes it do its work. Third, nested telecommunications software and other components needed to communicate with other machines (i.e., systems) are positioned to participate. Such components include software, telephony, and the Internet. The top layer is the most visible to people: applications such as word processing and social media platforms (e.g., Facebook). The schema has proven useful for decades to illustrate the relationships of one technology or component to another.

Now examine figure 6.2. It looks deceptively simple, but the implications for how computer designers worked quickly proved profound. The image suggests that transactions—activities, processing—begin by "feeding" a

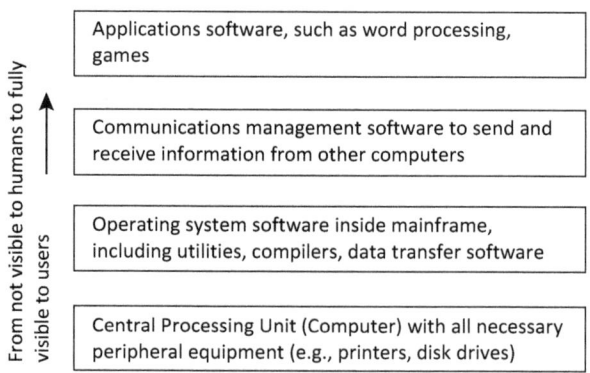

Figure 6.1   Schematic of a Computer System.

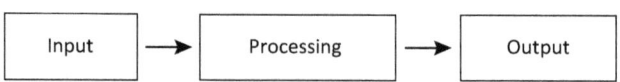

Figure 6.2   Von Neumann Architecture.

computer data and instructions (software) on what to do with it. In goes the information (usually referred to as data), where it is processed per instructions, with results then sent out to a storage device or some other machine that people can "read," such as a printed report or in human language format on a screen. So far, this is conceptually simple. This architecture prizes coordination among multiple pieces of hardware and software technology, hence viewing computing as an enclosed system (i.e., entity) and set of processes. But, of course, it is not simple, which is why the idea helped spawn the entire field of computer science, vendors, and millions of IT professionals who design, build, and use what are arguably the most complex machines humans have ever invented.

As technology and software evolved over time, innovations arrived incrementally in piecemeal forms. The speed at which data could move from input to processor might increase, but not necessarily at the same rate as it moved from computer to output. Often, the processor (i.e., computer) worked faster than the input devices could shovel data at it, so the processor might lie idle, waiting for the arrival of more facts to work on. Alternatively, the storage devices might be wicked fast but the channels (think paths, roads) moving information to and from these, depicted by the arrows, could be too slow. Thus, there was, and still is, a constant struggle to get everything optimized to handle speed and volumes in a coordinated manner. Optimization, too, keeps costs down for systems and their staff. To optimize and make effective all of that activity requires looking at an entire system holistically, constantly adapting one component or another to avoid damaging bits and pieces of the system, and always respecting the collaborative and coordinating features of the whole. That is a similar mindset that a biologist or cultural anthropologist embraces when looking at a jungle or village, identifying what makes the whole eco*system* work.

Figure 6.3 applies the concepts reflected in the two earlier figures to an ecosystem. In this instance, the major groups of participants (think components, bits, and pieces) of an ecosystem are positioned within their physical locality from on (or below) the ground to the top, in this instance above the trees. As for computer designers, this visualization assists scientists in making sure they include all participants and identifying where they fit into the ecosystem relative to each other. They then can use such a figure as a checklist of which pieces of the ecosystem they wish to study and to understand where their work fits into the larger schema of the jungle. Business employees in a large corporation do the same when they show a PowerPoint slide depicting their company's organization, highlighting in which box they reside.

Finally, a related method used by the IT community expands on this idea, demonstrated by figure 6.4. It is a standard, generic depiction of what they call *platforms*. The idea is that there is a hierarchy of broad-based technologies

152   *Chapter 6*

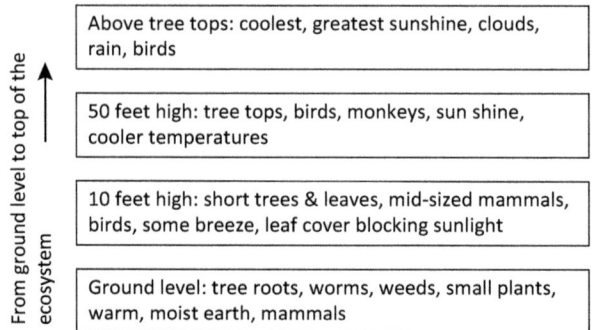

Figure 6.3   Schematic of a Jungle Ecosystem.

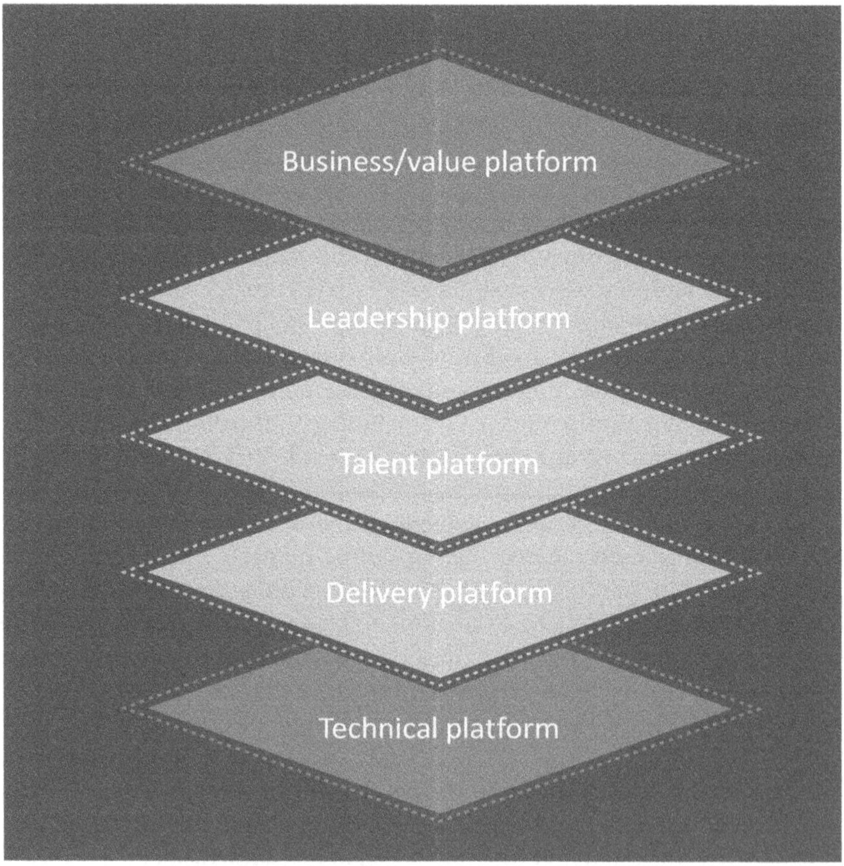

![Platform View figure with layers: Business/value platform, Leadership platform, Talent platform, Delivery platform, Technical platform]

Figure 6.4   Schematic of Platform View. This one illustrates the hierarchy of business values.

up to highly sophisticated or specific ones, uses, or functions. Think of each as a floor of a company's building, with each level home to a specific department. On each platform resides information or specific functions. Facebook sees itself as a platform upon which its users and advertisers carry on their functions. Think of a platform as an office or ballroom at a hotel used for conferences and wedding parties, theoretically agnostic as to what occurs on it. Some are dependent on lower ones to be useful (e.g., kitchen services at a hotel), others not so.

This conceptual schematic is not simple or casual. Today, for example, with massive amounts of misinformation flowing through Facebook with negative consequences for politics, society, and resistance to vaccinations, the company's executives resist taking responsibility for these consequences, arguing that the company is merely a platform on which other parties can do whatever they want, invoking freedom of speech pabulum. Regulators are pushing back, arguing that platform managers have a cultural and legal responsibility to ensure socially deemed inappropriate behavior does not take place on these platforms. Nonetheless, using the concept of platforms to position types of information in relation to each other works, so too in explaining who does what with information, and how one organizes discussions about it. All four figures represent familiar models for describing relationships among concepts, physical artifacts and environments, and actions. They prove useful for conceptualizing our thinking about information and its ecosystems.

Going through these four figures reinforces the notion that everything operates in some coordinated fashion with everything else, including information. Why do we care? Because a fundamental reason why information evolved was in response to expedients, such as searching for a new piece of scientific knowledge or making new goods, such as a medicine or a new chemically based product. All manner of people added these to their prior accumulations of information (often, too, called knowledge), existing procedures, and now new or modified ones based on new information. Thus, entire collections of facts are applied to continuously changing realities. Jobs were created and lost in the process for nearly 200 years at rates higher than before the mid-1800s.

WHAT ROLE DOES INDIVIDUAL HUMAN INTELLIGENCE PLAY? IQ AND INFORMATION

Lurking in the scattered corners of such disciplines as education, psychology, economics, and information studies are discussions about the effects of modern society on people. This discussion has a long history, but for our immediate concerns, we will focus on what took place since the start of the Industrial Revolution on humans. Independently of each other, they ponder similar questions, not the least of which involve human intelligence and, increasingly today, the rising capabilities of software. The subject is rife with controversy as it entwines with politics, discourses about races and social classes, and always, it seems, with society's controversies. But given all those risks to any discussion today about IQ (intelligence quotient), we cannot ignore its role in how scientists and others understand it vis-à-vis information if one is to determine how to appreciate the function of information. Discussing IQ issues is still fraught with emotion among experts one calls upon to rationally discuss IQ and information. It has a sorry history because IQ ratings and the application of statistics to explain cognitive variances in humans were often, to quote two experts on the history of information, a "tale of attempts to prove that social hierarchies rest on innate differences between people, whether differentiated by sex, race, or class."[48] They neatly summarized a large part of the IQ discussion problem by labeling it "the hubris of scientific racism."[49] It remains such a sensitive issue that most students of information and publishers avoid it.

In addition to its social and political difficulties, scientific knowledge about it also remains contentious because there is not a settled body of knowledge about the subject. But, information about personalities and mental conditions became both discovered and created information.[50] In psychology, for instance, naive perspectives continue to circulate, as two observers pointed out: "a person's intelligence is just how well the person performs on an intelligence test, nothing more or less."[51] While that quote is too simplistic, it suggests that because IQ tests measure something, a rigorous scientific approach is being applied. Nonetheless, the topic increasingly has to be confronted as our understanding of how humans engage with information as it expands. Situated at the core of the intelligence issue is one of many questions: Are people smarter today than, say, in 1800? If so, why? If true, and according to many tracking results of IQ tests given over the past 150 years this seems so, is it caused by education, demands of society, biological evolution of the brain, better nutrition, or access to more information and cognitively stimulating uses of it? In our age that so celebrates the quantitative, more than a century of IQ test results suggest strongly that in every decade children, in particular, scored incrementally higher than their parents or grandparents. However, if one took the rates of increases in IQ results tabulated, say, for the second half of the twentieth century just in the United States and ran the numbers backward to 1900, one would mathematically conclude that the nation was filled with many more stupid people and that the number who would have been considered cognitively sufficiently incompetent to place in asylums should have been a much larger proportion of society.

So something is wrong here. The statistics might be empirically (mathematically) correct, but we know from earlier research about agencies and organizations tracking cognitively challenged people that these individuals remained essentially the same percentage of society over the past two centuries (roughly 2 percent). Many readers probably have known people born around 1900–1910 who they knew were not low-IQ folks. They also probably appreciate that their grandparents could not all have been dimwitted, but rather that they were as smart and often as successful socially and economically as subsequent generations of their families.

My grandfather may have only completed 6 years of formal education, but he could add up in his head ten rows of numbers, each with ten numbers, just about as fast as a handheld calculator does today. I cannot do that, but I hold three post-secondary academic degrees. My mother was considered extremely intelligent by friends and family, graduated number one in her high school, but never went to college. Blame that circumstance on the economic impact on her family from the Great Depression. My father-in-law, a farmer, completed eight years of elementary school but turned out to be highly successful in his profession. Not a weed on his property, beloved by his entire county, and he managed to send both of his daughters to college. He also died debt free, quite an achievement for an American farmer in his time. We all have stories like that.

So, what is happening? Based on research done on IQ tests across the twentieth century, all of these individuals should have done terribly, or at least poorer than my sisters or me. I can assure you I am no genius (I saw my IQ scores from high school tests), nor a brilliant scholar. I also know that many of the people I interact with are much smarter than I am. That is not false modesty; rather, it is based on observation of their performance.

Careful studies by psychologists such as Alexander R. Luria and James R. Flynn, among others, provide an answer, not just a clue. Their findings are explicitly linked to the evolution of information and its use during the Industrial Revolution. Further, their research links directly to changing information and the evolution of human cognitive capabilities. This kind of work needs to be integrated more fully into our conceptions of the nature and role of information in modern society.

Alexander Luria (1902–1977) was a leading neuropsychologist. He studied Russian peasants in the 1920s, people who had little education and did not see the world through "scientific spectacles" (his term), unlike those who had four or more years of classroom education. In the United States and most of Western Europe, each subsequent generation stayed in school longer. Today, over a quarter of Americans aged twenty-five years or older have engaged in post-secondary education, such as colleges and universities. Luria observed that those who had gone through some formal education displayed different cognitive skills. Briefly summarized, recall that he was one of the first psychologists to observe that formal education, be it ever so brief or not, made it possible for people to think of theoretical possibilities, not just about immediate tangible problems to resolve or actions to take. People could move from just thinking about practical activities to broader considerations. These included moving from sensory responses to more abstract considerations, categorizing information, and considering consequences and possibilities beyond one's immediate experience, drawing logical inferences as well, "regardless of whether or not the content [information] of the premise forms a part of personal experiences."[52] Discursive thinking became more evident. Larger assumptions, hypotheses, and frameworks emanated as byproducts of formal education. As subsequent psychologists demonstrated, that changed relationship to logical reasoning was an important byproduct of the Second Industrial Revolution.

For understanding that next step, turn to James R. Flynn (1934–2020). He studied the question of why IQ test results rose throughout the twentieth century, raising two questions of what caused that rise and why. He analyzed all the major IQ tests used in the twentieth century. His observation: as people gained more education, that is to say, more exposure to such scientific norms as the ability to compare, embrace precision, form hypotheses, and engage in abstract thought, among others, the better they performed on such tests. By "better," he meant that people who took these tests based on an old edition that compared the individual to a cohort tested even twenty-five years earlier, which had less (or different) education, led to the latest test-takers scoring numerically higher and hence appeared more intelligent. The more a person's test results reflected the shared educational experience of the individual, the less likely they would *not* appear much cleverer. He measured numerically how such tests misled, as the cohorts to whom someone was being compared differed over time. As information and its influence on education and increasingly cognitively demanding work affected people over more than the past century, the more testing questions and comparisons of cohorts evolved, or, put in other words, mislead.[53] Cohort comparisons are roughly updated about every twenty-five years. So, a 14-year-old taking an IQ test in the 1970s might be compared to 14 year olds who took the test in the 1940s. The former would be expected to score a few points higher because of his or her additional education. On the other hand, if that individual took the same test just a year or two later, he or she might be compared to the scores of 14 year olds' scores of others of the 1970s. That child could be expected to have a lower score, that is to say, one more normal to those of his or her generation. Flynn reached the obvious conclusion: "Scoring children against obsolete norms created an absurd situation in which every one of America's fifty states claimed that their school-children were above the national average. School officials and parents were gratified."[54]

Other factors, too, were in play. For example, in the United States, Asian Americans are considered smarter than the population at large by many in the public and hence are admitted to more prestigious universities. They acquire attractive jobs in leading companies and universities. Flynn argued, with evidence, that their IQ levels were no greater than those of other Americans. However, Asian parents applied their social values by providing their children cognitively enriched and more challenging upbringings that valued study and hard work, thus improving their intellectual and later work productivity.[55] Flynn explains, perhaps unfairly to one ethnic group but with a point to make: "If an Irish lad qualifies for an elite university and his fiancée wants him to stay home, he may do so. A Chinese youth is likely to get a new fiancée."[56] It was a strong work ethic and extensive education that made Asian people seem more intelligent, not their cognitive muscles or genetics—brain—being better, that is, "smarter." For them, "high achievement preceded high IQ rather than the reverse."[57] Flynn was siding with those who advocated that one's environment affected their abilities.

There are many implications for the study of information when such findings are linked. For example, cognitively rich professions increased as a percentage of all employment during the Second Industrial Revolution. These included managerial, technical, academic, law, and accounting. These always required the highest IQs in the workforce, but over time, the IQs of these professions declined as the "average" lawyer, when compared to other lawyers, for example, had lower IQ scores, as explained in the previous paragraphs. Flynn cited data that showed that these professions collectively had mean IQs in 1950 of 111.5 but in 2000 had dropped to 110.5.[58] The latter were not less intelligent than their cohorts of the 1950s, just that the norms had changed. All cohorts were more similar in their education and possibly social positions. The newer group was armed with new or different information and tools that made them productive. Flynn demonstrated that "as elite occupations became less elite, that is, as the percentage of those who are managers, lawyers, and technicians rises, the IQ thresholds tend to fall."[59] Now inject AI into this situation, where some of the tasks and decision-making performed by these people are increasingly being automated, thanks directly to the combination of computers able to learn from prior experiences and to collect massive quantities of data with which to be informed, and one can see where this conversation could go.

Back to the anthropomorphic, as educational levels improve, say in Africa, Africans should score better on IQ tests than they do now, while individuals in "advanced" economies will level off or drop in comparison to earlier cohorts. Add to the mix that educators and the educated developed the language and norms of science in wide use over the past two centuries, and one begins to see links formed among education, scientific thinking, and more cognitively centered work, hence specializations by discipline and professions. All sit astride information.

One cannot be blamed if they conclude that this development constitutes a revolution in the human condition, less so by some genetically induced increase in native intelligence. If it were not so, either the human species would have been eliminated for lack of sufficient intelligence a long time ago, or we might still be stuck in some nomadic or primitive agricultural era. Flynn speculated that IQ levels may be leveling off, less because of AI or some genetic factor but rather because IQ's "rise" (his word) occurs "in tandem and boosts one another in a cycle of reciprocal causation."[60] In other words, humans would continue to do what they have always done: respond collectively to the realities of their time and place in order to survive and thrive. Other psychologists also recognized that some people are better capable of performing certain cognitive tasks than others, such as a heightened ability to learn and use mathematics essential in many hard sciences. Some individuals are more mechanically oriented than other people; still, others can better "connect the dots" in understanding social or political circumstances.[61] Most people optimize what they are told (or believe) they are good at, leveraging professions and information to their advantage. Their hunter-gatherer ancestors did the same.

One could conclude that we have tied up the whole conversation of IQ—intelligence—and its relation to information in a pretty package so that we can move on to a discussion of AI and its more fashionable current iteration, "generative AI."[62] Behind the scenes of research on IQs is a stream of investigations that seem to compromise the earlier work on information-based intelligence and even AI: emotional intelligence (EI). While space does not permit discussing it at length, be aware of it. EI concerns the ability of one to perceive, understand, use, and control (manage) emotions, both theirs and those of others. We all know people (mostly relatives) who are geniuses in doing that.[63] It was identified as a possible form of IQ in the 1950s and it is still embryonic in how it is understood.[64] The hunt is on for descriptions of its features and how best to measure it, but there is general acceptance that it probably exists. Students of EI build models of its characteristics, theories of how it develops, and consequences for improving social relations, say, among children and adults, in improving academic achievements, improved ability to negotiate, and ultimately, overall well-being in society and with friends and family, and to one's personal overall mental and physical health. If this discussion strays from the more specific definitions of information, group behaviors, and roles of computing and AI, it means the subject of how humans and information continue to transform, and along the way, individual IQ.[65]

But we are still left with one IQ-related issue—the role of group intelligence, or what some call "swarm behavior," such as the collaborative work done by bees and ants and now increasingly by people through their civilizations with "crowdsourcing." Are such behaviors manifestations of group intelligence? Is it a capability in which extensive collaboration is made possible among diverse masters of bodies of information, such as lawyers and scientists, biologists and computer engineers, historians and sociologists, applied to circumstances requiring deep knowledge of such diverse bodies of facts not practical for one individual to master?[66] People have collaborated for centuries; we have only to think of the diverse authors of the *Encyclopédie*, published between 1751 and 1772. Team-based research became standard practice in university and corporate laboratories by the end of the 1920s. Computing reinforced that model of work and application of multiple bodies of information to projects by the 1980s. In the 1990s, companies began to pose complex problems on the Internet, offering to award cash to whoever came up with a resolution. Retired chemists, physicists, hobbyists, and medical experts participated both as individuals and in groups. It worked. Bees and ants display a cumulative intelligence, created by accumulating into a collaborative initiative bits and pieces of capabilities that by themselves are not effective, but when combined make it possible to build complex hives. It is why these and other creatures have been studied, to learn what collective intelligence looks like.

Humans behave with collective intelligence when crowdsourcing, also by sharing databases of data and facts. Like colonies of ants, the Internet has no central authority, no "ant-in-charge." One could argue that it is not anthropomorphic, as we mortals are reluctant to give up our individual claims to intelligence. However, as one expert on Internet functions bluntly put it:

> intelligence is intelligence no matter what form it comes in, just as ice and steam are both just water at heart. Swarm intelligence is merely a highly transitional state built upon the compounded contribution of many, many minds, whether each is as insignificant as that of a single ant or as Earth-shattering as Einstein's huge intellect.[67]

Undergirding this Internet-based activity is the combination of computing's ability to gather much data and allow people to coordinate the application of all manner of information, most of which was created in the past two hundred years, most intensively so in the past several decades. This development may represent the next chapter in the history of how information is created, used, and changed. George Dyson, a historian and commentator about technology, might have leaped ahead just beyond what is happening today, but essentially right when he argued that "machines began taking the side of nature, and nature began taking the side of machines. Humans were still in the loop but no longer in control."[68] That was his way of introducing the next, and probably most aggressive collector and user of information the world has ever seen—AI as part of computing, discussed in the next chapter.

AS A WAY OF UNDERSTANDING INFORMATION— THE ROLE OF CONTEXT

People have understood for millennia that a fact floating about untethered to a circumstance or reality is not as well understood (or appreciated) as those ensconced with other facts or appreciation for its own intrinsic messages. There are exceptions, of course, such as in mathematics: 2 x 2 will always equal four. But explaining why librarians took steps to implement cataloging systems in the nineteenth and twentieth centuries makes more sense if we understand the problems they were trying to solve or, more broadly, what society expected of them. Answering the question "Why?" has long served as a method to link facts to relevant context. So intelligence and issues concerning the organization of information are also improved by understanding the circumstances in which it is applied. This is a point made several times in this book through discussions about context. Before concluding this chapter, it is crucial to revisit context, as it will become a topic of greater importance as humanity increases its various overt uses of organized information, especially using AI.

Recall that context is about understanding the meaning and role of a fact in an ecosystem. We must understand how it is applied, or remain prisoners of its theoretical and academic definitions. An astronomer studying planets and galaxies will deal with a new fact about black holes differently than someone ignorant of astronomy. Heuristic thinking practices are variations of relying on context to give fuller meaning to specific facts.[69] Experts and people who are considered wise filter facts through context. Sometimes this act is seen as doing so through our biases, or what else we know or have experienced—fair enough. Teachers and managers use that mental process to promote decision-making based on facts that they see as "better." That behavior is central to modern notions of *nudge*, in which someone wants an individual to default to a desired conclusion about the meaning of a fact, hence taking actions based on a piece of information situated in some context.[70]

Less controversy exists about the value of context—some call it experience or wisdom—than in the study of biases. It is probably accurate to declare that as we age, or as we learn more about a topic, the more we value context. Its value has increased over time but has become more illusive to acquire. There is more information to master today about any subject than one needed, say, a century earlier. So, one increasingly finds context within a specialized body of information, while in prior times people learned about different collections of information from which they could extract relevant context. As information increases, the amount of time one can devote to learning from other specialties to broaden one's contextual understanding of a fact diminishes. That reality perhaps explains why a thirty-year-old professor of astronomy might be ignorant of much history of astronomy but, fifty years later, would have finally accumulated enough knowledge of the subject to make her current research better appreciated.

One byproduct of specialization has been the resurgent celebration of the generalist, the person whose knowledge of multiple subjects may be, as the expression goes, an inch deep but a foot wide, as opposed to that of specialists whose understanding is a foot deep but only an inch wide. The latter was encouraged. For example, a young professor eager to obtain tenure was—is—expected to conduct research in their specialty and to publish only in the key journals of their discipline. With so much to master, one can hardly blame a university or employer for laying down such strictures. But the variety of disciplines and professions increased so much over the past two centuries that walls went up around bodies of information that need to be breached. There were always people who did that, and often defenders of the liberal arts argue that they teach people to do that. Historians are famous for making similar arguments. Both are frequently correct.[71] In the late twentieth century, it was becoming clear that there was value in drawing insights from multiple disciplines, which is why, for example, multidisciplinary teams increasingly worked on complex problems.

Earlier chapters described efforts to somehow merge discrete bodies of data and information into composite wholes larger than mere mortals could integrate. Algorithms that Google uses to scan hundreds of disciplines for facts relevant to someone with an immediate need represent one example. Another is the development of hardware and software that could accumulate Big Data. Imposing statistical practices on these new collections of information made possible new insights about information, set within contexts (parameters). How the public has been informed about COVID-19's spread and evolution is a visible current example of this capability at work. For instance, instead of just relying on scientific knowledge about the spread of this coronavirus, one could also overlay GPS data from smartphones to track its spread from one place to another with information about where millions of people traveled to and from, say, a super spreader event.[72]

Specialization, as reflected in the growing number of technical papers and patents over the past century, has done more than consume additional time from those generating new information. Studies of the innovative qualities of all this new information are beginning to demonstrate that, as a whole, they are serving up fewer innovations and less disruptive findings that augment one's understanding of the world. Put another way, there is growing evidence that innovation is actually slowing as more information is published. Creators of significant new information do not seem to have enough time to create something as novel and disruptive as, say, modern perspectives about DNA or the creation of computer chips, both from the 1950s. Rather, people build on prior innovations and breakthroughs by incrementally building and writing on the backs of those earlier innovative researchers, citing ideas and text with which they are already familiar. Scientists continuously rely on ever-narrower bodies of information, while in the humanities, some scholars are arguing that more information eventually leads to new discoveries and information-based breakthroughs. Evidence suggests that innovations and breakthroughs in new information continue to come but at a regular pace as in prior decades, even though the amount of new information increases, as measured by volumes of publications and patents.[73] So, connecting the dots between disparate bodies of information is not happening as quickly as the growing amount of information would suggest, hence the attraction of context, which is seen as connecting intellectual tissue.

Context has long also been valued as a predictive tool. For example, based on my nearly forty years of experience at IBM, I could expect that a specific computing product would do well (or not) within a certain period of time. An executive trying to make a decision on whether to invest in the development of that product might value this unquantified, perhaps even indefensible, assertion if my contextual expertise is respected. Often, such a contextual perspective trumps empirically based alternatives. Recall our discussion of Flynn's IQ studies where statistical explanations of the growth of intelligence, hence the logically lower performance of earlier generations, could be explained away with the application of context.

Enter AI. Today, the aspiration is for AI to augment the capabilities and priorities of the human mind. Whether this technology eventually surpasses that of people and sets its own agenda is only interesting reading because it has yet to achieve its immediate intended destiny. But AI engages two rapidly expanding capabilities: its natural language processing and "deep learning." Put another way, with the ability of computing to process ever-larger collections of data, software is increasing the ability of AI to identify patterns, hence augmenting the accuracy of predictions. Pattern recognition can theoretically be applied to any collection of data or theme, and it is doing quite well in this endeavor so far. Humans do that too, albeit with less data but with far more sophistication than software, also so far. Pattern recognition is at the heart of contextual thinking. In many societies, there has long existed the belief that when a person dies, so too does a library. They are talking about experience and collections of information (e.g., about a family's history), but most frequently about wisdom, which is the filtering of specific and eclectic facts through the lens of context. AI promises (again) to play a growing role in the development of context and possibly at some distant time even better than what humans can do today. Scaled up, when matched with predictability, the nature of information may go through another fundamental shift beyond what we have witnessed over the past two centuries.

FINALLY, SO WHAT?

Momentarily leave aside the cyborgs, AI, and smart robots, as we have more immediate issues to address. Fundamentally, across the research that went into the three volumes of *All the Facts, Birth of Modern Facts* and this book, several insights surfaced. First, the closer one approached the present, the more information was created by humans, and increasingly now, too, by sensors collaborating with computing and the Internet. Second, over time, new information became more specialized, posing the chicken-or-egg problem of which came first, specialized professions or specialized information? Either way, we have both. Third, humans became increasingly reliant on organized bodies of information with which to go about their lives. Fourth, many of the features and behaviors of scientific methods and knowledge enhanced the way humans cognitively dealt with the world. What one learned from one body of information, they applied in another domain.

A leading observer of scientific behavior, Jim Al-Khalili, observed that "you will also find physicists applying their logical, numeric, and problem-solving skills outside of science in professions ranging from politics to finance."[74] The fact is, this behavior is in evidence across all disciplines and skills. That may have been one of the most significant anthropomorphic consequences of more information than many others that focus on economic development, improved quality of life, and medical marvels. This is because, without the changes in human cognitive behavior, none of this might have happened. In other words, the ecosystems humans lived in changed how they processed information. Global ecosystems are rapidly forming a thick wrapping of a community of information, bigger and more comprehensive than what one person can accumulate. Without it, no single human could build a car, fly an aircraft, or study the worldwide effects of a virus. Billions of humans have come to see their world through "scientific spectacles" in combination with large bodies of information housed outside their bodies. The changes in information over just the past two centuries have altered how humans use facts for many more to come. If ever there was a revolution in human history, that might be the biggest.

Of course, we would expect—and find—that the debate continues about the future of information and how humans fit into the discussion and the realities discovered. As one thoughtful student of information put it, Caleb

Scharf, when describing the ascent of information as both an issue and an agent affecting reality: information "isn't just a way to probe the fundamentals of nature; it may be part of the fundamentals." Consequently, humans are "becoming increasingly entwined with the fabric of the universe—as pieces of manipulated matter and energy," meaning that people, as "living things are fully committed to the universal drive toward that future ocean of unchanging, equilibrated space-time. It is as if we popped out of the vacuum as a temporary fluctuation of energy, and we've been clawing our way back ever since."[75] Our awareness of the nature of information and increasingly of its effects on us, our physical world, and all that inhabit the universe must be seen as profoundly central to what we learn and do next. Much has changed in human perspectives about everything over the past two centuries, and that changed view came about by our growing understanding and acceptance of the role of information.

Permeating all the chapters of this and the previous study (*Birth of Modern Facts*) were themes of increasing specialization and quantity of information. Also bubbling forth in the volume you are reading is the rapid and muscular intrusion of the Internet and computing into what had been a very labor- and human-intensive activity for centuries. Given the growing evidence since the 1950s of IT crossing disciplinary borders, are we nearing a time when all information is essentially one big soup? Already, too, we are seeing expertise and credentialing decline as people go to the Internet to get, say, a quick YouTube discourse on some historical topic produced by someone not credentialed in a traditional way to discuss the topic. Will that be a new normal behavior in the use of information?

People are increasingly relying on the Web for information not curated by traditional standards, so they get bits and pieces of information, probably not in context unless the software that answers a question chooses to either offer it up as part of an answer or factor it into the creation of a response. Take history, for example, where increasingly people think they can self-learn about a topic, with the possible consequence that majoring in history in college is no longer necessary. Consequently, history departments have begun to shrink, while academic historians are outgunned by almost "anyone." To be sure, hard data is presented constantly that newly graduated students who majored in the humanities begin with lower salaries than those who studied engineering or the sciences. However, over the course of their work lives, both end up earning more or less the same and achieving similar success in similar careers. While professional historians accurately see such developments as existential threats, the public demonstrates a strong appetite for historical information, as measured by their visitations to myriad websites and books written by journalists and "amateur" historians. As one student of the trend concluded, expertise is an "obstacle, not an asset," especially on the

Internet.[76] Users increasingly determine upon whom to confer accreditation, not universities and professional associations. This behavior is not limited to history. So what? Well, will information become better and more useful over time? We shall see.

We have discussed a broad range of issues concerning today's form and use of information, or facts, if one prefers. With that very broad context of understanding, we can turn to the next major challenge and opportunity facing humankind: AI. We must resist the hype, even the hysteria and fear all over the Internet and proffered by thoughtful observers, because it is such an important topic. That is why the last chapter is devoted to this subject.

Chapter 7

The Special Issue of Artificial Intelligence

We need to know—and, of course, we must also understand to make productive use of—a great many truths.

—Harry G. Frankfurt[1]

Prior to 2020–2022, the only people who understood that AI was about to experience a step change in capability were computer scientists, and even in that rarefied world of several million people worldwide, only a subset knew specifically what would soon unfold. As noted in previous chapters, computing evolved substantially and rapidly throughout the past three-quarters of a century, with periodic spurts in innovation that made information handling easier, different, and less expensive. Artificial intelligence (AI) had bumped along with much interest in the computing community but with only highly specialized uses and not as dramatically as, say, the invention of computer chips or the Internet. But that circumstance appears now to be changing sufficiently and quickly enough that, to complete our discussion about how information is changing and shaping the world, we need to focus specifically on AI. It is the next frontier in information that will undoubtedly affect everything discussed earlier in this book deep into the 2030s. As this chapter was being written in 2024, it seemed almost every computing publication, most book publishers, and all news media outlets were extensively commenting on AI to an extent never witnessed in the previous seventy-five years of AI history. Then in October 2024 five Nobel Prizes went to experts on AI. One of them, Geoffrey E. Hinton, a co-recipient of the prize in physics commenting on the significance of AI informed the world that "It will be comparable with the Industrial Revolution," adding that "Instead of exceeding people in physical strength, it's going to exceed people in intellectual ability. We have no experience of what it's like to have things smarter than us."

The hype, both accurate and misinformed, is massive. Regulators, other government officials, and legislators are attempting to understand its implications. They are trying to stay up-to-date with their policies and laws as the technology keeps evolving at a pace challenging the capabilities of institutional responses in both the public and private sectors. Politicians, military personnel, political scientists, and media commentators also fear that fake information will be pumped through the new AI in highly believable forms to confuse voters and officials; the same with advertising. Parents are concerned children will be targeted, too, and that college students will use AI to write all their term papers. Yet others see the new AI as tools to help educate adults on how to use various products, to interact with medical providers, and to help them improve their Internet searches far beyond what they experience today. All those many hundreds of books being published today about AI will all be out of date faster than most publishers can recoup their costs to produce. It is also reasonable to expect that what readers understand about AI will be too, including what is discussed in this chapter. But, we must begin somewhere to keep up with this new round of technological changes. So, we need to understand today's AI to a degree one might not have considered necessary as recently as, say, at the start of the COVID-19 pandemic in 2020.

We proceed by suggesting how to understand the role of AI since it is now changing sharply and rapidly making its way into daily lives of people and organizations. We then discuss recent developments by picking up from where we left off in the first chapter and conclude with a brief discussion of robotics, which are also now going through a transformation of their own. Our purpose is less to explain the specifics of the technology and its information and more to equip the reader with ways to process what they are learning about it.

A WAY OF THINKING ABOUT AI

American and Western societies have long paid attention to issues immediately in front of them, often meaning those affecting jobs this year and their politics shaping up for the next election. People's attention spans are increasingly influenced by a growing reliance on social media and fragmented news sources for information over earlier dependence on such trusted sources as mainstream news media and what they learned in school. Let's be candid: their attention spans are becoming shorter, the "elevator answer" more appreciated, and less so longer-term concentrations needed to read books. All of these practices discourage "deep dives" into large bodies of information. Other factors are in play, too. For instance, declining attendance at religious services by younger generations in the Western world and, on the extreme political and religious right, less reliance on scientific and other factual sources with which to shape their worldviews, all of which contribute to a messy society-wide potpourri of views about information. It is also becoming increasingly evident that entire societies are less likely today to have a commonly shared, collective knowledge of the same facts. What they learn about information and other topics is increasingly being supplied via computing in fragmented variations. That is how we get to issues of AI because each of these trends will undoubtedly be extended by the next round of software. The use of AI has already started that process.

Without their consent, the public's world, America's in particular, has quietly experienced aspects of their lives being transformed through expanded use of AI by social media, government intelligence and law enforcement agencies, political parties, and the one source frequently talked about: social media and Amazon's appropriation of private information since the early 2000s. After seven decades of development, AI has evolved to a point where it is used in highly diverse activities. AI is a topic deeply embedded in many discussions one might want to have about the modern role of information. Information about AI and its influence on information is now extensive. Two manifestations of immediate interest are the way organizations use it to influence the thinking of over 3 billion people, nurtured by information sources they rely upon, and robotics, which, too, are entering a new era of appropriation. The first affects views on all manner of subjects, the second the potential for jobs transformed or replaced by machines or software. Both have been occurring for nearly two decades, but the public is barely aware of that. This is not the place to unveil that reality; others have done so.[2] Step over that body of research to the present when AI and robotics are sweeping into our lives at probably the same speed as computing did between the 1970s and the early 2000s.

Anyone who was an adult in those earlier decades in the United States, Japan, or Western Europe should be able to recollect how much and how quickly things changed: microwaves in cooking, online systems and databases forcing people to fill out online forms and rely on computerized information, secretaries becoming a nearly extinct species, and travel agents by the end of the 1990s, ATMs becoming common and by the early 2000s nearly ubiquitous online banking. Cars became highly electrified with software controlling pollution, braking, and lights. Today, electric and autonomous cars are already driving around, thanks to AI connected to the Internet. It has been half a half century since a human painted a newly manufactured car in an automotive factory—mortals are too sloppy compared to a robot and get sick, go on vacation, or strike for higher salaries. We are probably with AI and robotics today where we were with computing in the late 1960s. AI has been used to target people with tailored advertisements since the early 2000s, based on their prior online searches and purchases. AI is even more obvious in our lives today.

Historical experience suggests that we have been here before and not just in the 1970s: in the 1830–1850s with the implications of steam on worldviews; increased speed of travel, and changes in and transforming jobs; again in the 1850–1900s with the infusion of electricity into communications, machines, and lifestyles (thanks to the early "killer app"—lights), and additional changes and losses of jobs; and, of course, the arrival of computers in the 1950–1960s. We have encountered disinformation, too, for centuries. Much of it today nests in the Internet.[3] So because of these realities, it is important to think about AI with that historical perspective in mind as it helps us to understand what is currently happening. Simply put, one can be informed by what historians have learned about how information has—and is—changing.

Despite advancements brought forth in all manner of life's activities by way of technological innovations, progress in medical practices and medicines, and the general widespread application of scientific practices that brought forth so much information, too many people simply believe what someone posted on the Internet or what some media, publishing, or news outlet presented without scrutinizing what they read. That is the fake facts conversation held in chapter 5. But there is yet another aspect of information to consider related to AI: hype. Some call it marketing, others advertising. It is an old blemish on information, business practices, and technologies that appears periodically. It surfaces as elegantly as in prior centuries and not always as sinister as hype would have it. A brief case in point involves the normally respected IBM Corporation and AI. It offers a cautionary warning.

Like such other firms as Amazon, Microsoft, and Google, IBM in the 2010s began developing plans for converting the increasing learning and data-gathering capabilities of AI systems into revenue-generating offerings. It wisely wanted to package AI capabilities into software tools and services that aligned with the specific information needs of individual industries, such as AI for insurance, AI for banking, AI for supply chain management, AI for law enforcement, and so forth. One promising area was healthcare, using AI to advise doctors on what medical treatments to use with individual patients by surveying the vast medical literature to provide probabilities of what the diagnosis should be and of success among optional treatments. It was an application for AI that seemed ideal and, obviously, if it was going to help doctors, a welcome development. IBM had already branded its AI capabilities by giving it the name Watson, after its long-serving CEO of the first half of the twentieth century, Thomas J. Watson, Sr. (1872–1956).

IBM's marketing machine went full steam ahead touting the capabilities of Watson and its medical prowess. IBM spent vast sums on its development, calling it "the future of knowing." However, marketing and a sales force dedicated to Watson oversold it to the medical community and to the public. Technically, AI could not yet deliver what the company was promising, but its executives, the IT industry, and the general media fell in love with the promise. It proved unable to provide doctors with the convincing, confidence-required advice they needed. It was not yet sophisticated enough. But marketing and senior management, too, may not have fully appreciated that reality, although they believed they did.[4] By 2019–2020, it had become obvious that it was more hype than practical, and so IBM pulled back its offering, reducing its commitment of people and other resources to that. IBM management became more humble and less exuberant about AI's capabilities. IBM recognized that before AI was ready to be the next great thing, much more work would need to be done, at least for some uses, as in medicine.[5]

Also lurking in the shadows that historians of the medical profession could have pointed out is the nearly notorious resistance that the medical community displayed when it came to new technologies. A Harvard teaching doctor in 1871 admitted to not knowing how to use a microscope, doctors and their hospitals resisted using computers in the 1960s and 1970s, today's "secure" email communications with our doctors are well over a decade behind the times, some clinics still accept faxes, and so it goes on. So, a lesson for IBM and other firms—the reader too—is that one should take into account the potential user of AI's attitude toward innovations of any kind when reading about all these marvelous new toys and how they will transform everything. They will change much, but like planting flowers, we need to also pay attention to the quality of the soil in which these are to flower.

IBM's "moon shot," as its CEO, Virginia Rometty, called it, assisted in the hype and hubris with statements such as, "Our moon shot will be the impact we have on health care. I'm absolutely positive about it."[6] This is not the place to debate whether she lied, got too excited about promoting it, or did not understand what it really could and could not do—historians will eventually reveal her intentions—nonetheless, she participated in an industry-wide touting of AI. But, as doctors pointed out, trying to apply AI to cancer "turned out to be really, really, hard." Doctors and IBM technologists had difficulty understanding each other's worldviews and so could not even agree on how to use AI as a diagnostic tool. That the company learned a great deal is beside the point, because the negative press masked significant progress that it had made, notably in natural language tasks such as in identifying places and people, then communicating with them in more human-like language.[7] These developments were already paying dividends that the health care initiative did not.

Computer vendors always seem to allow their marketing and advertising to get ahead of what their products were capable of doing. "Office of the future" was a 1960–1970s example, so too were distributed databases in the 1970s and 1980s, and "Smart Cities" in the 1990s. It was not just IBM at fault, but most of them, including their industry media. When I worked at IBM (1970–2010s) and saw something new being hyped, I said to myself, "OK, that will come in 4 to 5 years." I do not believe I was wrong in my forecasts; such an attitude led to some heated debates with customers who wanted the new offering before I felt it was ready for them.[8] Was it sinister lying or good old-fashioned hubris by the IT industry or IBM? Given the size of the company, hence the distance between marketing professionals and the engineers doing the hard work of inventing AI, I concluded neither was able to level set the other in sufficiently grounded reality. Each was doing their job. We, on the outside, were left not fully understanding. But the lesson was clear: think about news of the future of IT and information as overstated, displaying a penchant for being too enthusiastic when compared to the realities sheltered behind closed doors.

Already, students of AI and information had begun opining on lessons to be learned. One insightful example is worth quoting at length:

> Most of the first half century of artificial intelligence focused on combining logic with knowledge hard-coded into machines. Data collected from everyday activities was hardly the focus: it paled in prestige net to logic. In the last five years or so [2018–2023, ed. note], artificial intelligence and machine learning have begun to be used synonymously; it's a powerful thought-exercise to remember that it didn't have to be this way.[9]

Just learning from data defied conventions of scientific approaches; rules about how something worked were being ignored. James Bridle, a British-based artist and observer of the technology scene, proved a harsher critic, writing before the post-2022 surge in conversations about AI:

> we find ourselves in an intellectual, ontological dead end today. The primary method we have for evaluating the world—more data—is faltering. It's failing to account for complex, human-driven systems, and its failure is becoming obvious—not least because we've built a vast, planet-spanning information-sharing system for making it obvious to us.[10]

Old discussions about the power of logic (indirectly, too, scientific methods) and mountains of facts—data—had not gone away when a new chapter in AI's history seemed to burst on the scene. So what was all the buzz about the "new AI"?

THE ARRIVAL OF A NEW GENERATION OF AI

News and social media, scholars, and management went, to use a phrase, "crazy," beginning in November 2022 when a new AI software tool, called ChatGPT, became available. It produced text and images based on what users requested. Unlike, say, a Google search, which produces all those millions of links that one can go to for possible answers to questions, this one prepares answers in natural human language sentences and paragraphs. It can write essays, poetry, text in a Shakespearean style (if you want that), and even pass law and medical examinations. Known as "generative" software, it promised to take searches and aggregations of information to a new level, allowing students to have software write (generate) their assigned reports (instantly sounding alarms for teachers), creating fake news (a concern of experts), and augmenting the collection of ideas and facts for researchers. It sometimes invented facts and citations, so it was not perfect.

A storm of controversy immediately erupted, not experienced since the introduction of smart phones in 2007–2008, or even earlier with the arrival of social media. Deep into 2023, it was being hyped as the next major transformative innovation that could augment human knowledge, too, while its critics were almost as hyperbolic in their forecasting of doom and gloom.[11] Google quickly introduced its own version, while Microsoft announced it was investing billions of dollars in this new round of AI. But, as with prior IT tools, we are too early in the life of this new round of software to sufficiently understand how it will emerge and affect people, despite rapid and tangible improvements over prior forms of AI. For the moment, hype, speculations, and hubris are drawing "everyone's" attention.

So we continue to live in an age of natural exuberance, not simply awash in fake news. That means one should be wary of AI prognostications that it will solve all problems (probably many but not immediately) or, at the other extreme, that lawyers, accountants, newspaper reporters, and so many other jobs will be taken over by AI, or that all students will have computers write their papers and take their examinations. Much will happen, but not next year. Good work is underway, but not necessarily as advanced as we are often led to believe. AI experts candidly admit to that, but more often that the future of AI will be even greater. They are as excited about this possibility as those in biology are about how genetics and DNA investigations will lead to new vaccines and cures for serious human ailments.

Meanwhile, AI software is collecting data, learning from it, and consequently changing forms of information. Such behavior is almost beginning to emulate what humans did with information over the previous two hundred years, including creating controversies along the way. Sources of much online information are suing AI vendors for royalties to use their data, and regulators are considering data privacy legislation. Workers, of course, are again concerned about their certainty that many will lose their jobs or have them so changed as to obsolete their hard-earned skills. Worldwide, workers in many countries are opining about the implications in ways they have about earlier forms of computing. In a global survey conducted at the end of 2023, only 17 percent of Americans and 20 percent of Europeans thought AI would improve workplace productivity, while at the other extreme with positive feelings, 67 percent of surveyed Indians thought it would improve their work, and similarly, 65 percent in Indonesia, while Mexicans came in at 46 percent. Of course, the earlier one uses a new tool, the quicker they learn to improve its usefulness as sources of productivity and creativity. So far, productivity improves when generative AI is used to check the quality and accuracy of one's work or to quickly improve communications.[12]

Thus far, and probably for a long time, how AI collects and uses information will be determined by humans. Besides the technical limitations of other alternatives, AI lacks intentionality. As two experts pointed out: "What they [AI] can't do is read your mind to figure out your intentions—your goals and desires and your understanding about how to satisfy them—and then make those intentions their own in order to arrive at novel suggestions."[13] Yet, intentions drive a great deal of what the human mind does with the collection and use of information. Humans also can share intentions with other people through their use of language, causing groups of people to collaborate with a common purpose.

So what do we know so far that would be helpful for us mortals attempting to use this technology? First, its software can learn accurate and inaccurate facts and spam, even though it does not yet have the capability of discerning the difference. Performing that judgmental function is proving to be complicated but has to be resolved if, for example, we are to use AI to determine whether to offer a potential homeowner a mortgage or provide medical advice that, if taken could harm a patient or compromise the credibility of a doctor. European regulators are already insisting that better decision-making capabilities be developed. Second, not all errors are equal. Political spam leading someone to vote for the "wrong" candidate is not going to destroy democracy, but giving a doctor incorrect diagnostic data and recommendations could cost a patient their life. Cancer has been the subject of much AI work over the past three decades and so is an excellent example many will want to focus on with this new round of AI.

But back to some historical perspective to help: Some IBM scientists think AI has gone through three phases so far. Each phase began with optimism, excitement, and hope but ended with disappointment or hibernation. The first involved the use of rules to guide decisions: If your income is "x" and the amount you want to borrow is a percent of "x" then yes, offer the mortgage, or decline to approve the loan. It left out personal character and other considerations. This was known as a "symbolic" or "logical" form of AI. A second generation of programs partially mimicked the way the human brain functioned, relying largely on the findings of biologists. It seemed right to pursue, but the hardware and software were not yet up to the task—still are not—but making progress.[14]

Then along came "Big Data," discussed earlier in this book but warranting further attention. While not viewed as part of AI by its developers, its functions are essential to what is emerging today in AI. However, computer experts have learned from each phase that informs what they are developing now. Briefly put, training an AI system requires massive amounts of data; teaching an AI system to identify and recognize patterns also requires also massive quantities of computing far beyond what most systems could provide (hence the interest in quantum computing that holds the potential for providing that horsepower); changes in algorithms (software instructions) are pushing forward today's AI. These combine features of earlier phases of AI, or as one expert put it, "AI is now being effectively harnessed to improve AI."[15] That is the learning function that one hears about in almost every article and news report about today's generative AI. Finally, companies, government agencies, and universities are hiring technically trained people in sufficient numbers to create a new generation of IT tools and who are learning how to apply them.

In 2022 and 2023, media were filled with reports of documents being prepared by this new technology. Then, in February 2024, the public learned of a new version of AI, called Sora. It could create realistic videos based on text, such as: "create a video of a well dressed woman walking down a fashionable street in Tokyo." It looks exactly like that and is created in seconds. Of course, that means one could create a video filled with misinformation, but also for teaching facts, demonstrating how to do something, or for more compelling advertisements. Actors and movie producers had already sounded the alarm that computers could put them out of business. College students have been reported as enthusiastic about learning how to use such tools as part of equipping themselves with attractive work skills, while their young siblings were quietly experimenting with it to the horror of their elementary and high school teachers.[16] It is still early, so, too, fears of existing jobs being replaced are widespread, while people recognized that new jobs would be created as well.

Students and young workers, such as the Gen Z generation, however, welcomed these new forms of AI. Their behavior is consistent with how earlier forms of AI were perceived.

To sum up: the use of AI is beginning to reveal information—data if you will—and patterns of relationships among them more easily than in earlier computer eras. Presenting graphical visualizations of such mountains of relevant data is increasing our understanding of issues and realities. As two information scientists explain, AI in combination with visualization and narrative storytelling reveals "information that is otherwise difficult to perceive," thus offering the promise of improving decision-making and reducing errors, "not only in the *explanatory* phase of communicating results to an audience, but also in the *explanatory* phase in which patterns are first discovered."[17]

Parallel to the general public's encounters with AI is the experience faced by ever-more specialized academic and professional disciplines, discussed earlier in this book. To repeat: as more information and understanding of a topic developed, the more difficult it became for a specialist to take into consideration data and facts being developed in other disciplines that might be useful, indeed influential on their own work. This has been a known problem since the 1700s. AI was—and is—intended to take into account findings and facts from multiple disciplines. It is an old dream that, in its most modern form, began with Operations Research (OR) during World War II. Simply put, OR was about integrating findings from multiple disciplines, often through the use of mathematics, although hard scientists suspicious that it was some sort of soft, less-than-empirical set of social science practices. OR was accused of being the sin of "interdisciplinarity," all in support of more "scientific" decision-making, particularly by the U.S. military. It proved successful as a collection of practices throughout the Cold War era and laid the foundation for subsequent "expert systems" and AI work. OR never became its own discipline, such as physics or chemistry, but it did assist in understanding and dealing with problems, offering up theories and practical recommendations that transcended academic disciplines. The thoughtful Harvey J. Graff neatly summarized OR's contribution, calling it "an exemplary showcase for the fusion of scientific methodologies, practical questions, and systematic inquiry."[18] I like to think of it as a nearly pre-computer era parent of the concepts behind AI.

We are now at a point where the public at large can begin to directly experiment and learn from generative AI. Instructional software, videos, and publications are apparently "everywhere."[19]

WHAT ABOUT ROBOTICS?

The same broad experiences with AI apply to robots, even though the latter have been functioning in practical ways since the 1970s. The COVID-19 pandemic caused many job shortages around the world, and the ones seemingly, if one is to believe media accounts, are being taken over by robots equipped with AI. These jobs include servers in restaurants; hotel and rental check in with humanoid robotics, among other customer-facing jobs; and less visible uses, such as in automated trucking. Just to be clear, "humanoid robots" means robots that look like metal people with heads, torsos, arms, legs, and fingers—the latter being some of the most complicated components of a robot. Humanoid-shaped robots can fit onto production lines in factories already designed to be populated with humans who have arms, legs, fingers, and torsos, so do not have to look like earlier robotic devices that might have looked like a metal box with one "arm" that looked more like part of a bucket device used by tree trimmers or electrical utility workers. One can expect to see more humanoid robotics, discussed further later.

But first, while early examples of robotic functions exist, less sophisticated robots that do not look humanoid are more the norm. These include such devices as automotive painting machines, which repeat their functions over and over again and look more like one or more large arms. That painting function does not require advanced AI; it has been around for decades. Sensational stories of bug-sized robots, others that can climb stairs or cross terrain without eyesight, or can serve as recruits for armies as digital soldiers make for great copy, and there are pilot programs for each, but they are not yet in wide use. It often takes decades to go from crude experimental projects to widely used reliable technologies. Like AI, robotics is not immune to this reality. The technology is complicated, and their builders are only just now learning from the movements and decisions made by non-human living creatures, such as birds and bugs.

If humanoid robots are to be effective, their creators will have to draw upon collections of information they normally do not consult. They want to increase the intelligence of the humanoids and are learning to tap into the work of neuroscientists, anatomists, and computer scientists who specialize in machine learning to learn how to do that. Those who study bugs (specifically flies), rodents, and even babies are learning that intelligence is distributed in a body needed for crawling, flying, swimming, and performing other physical activities. Scientists label such capabilities (knowhow on how to perform such functions) as "embodied cognition." Those readers who play golf, ski, and swim, for example, think of this as "muscle memory." That kind of intelligence has to be inserted into the humanoids, not just the ability to communicate in human languages or reason in some anthropomorphic fashion. As one student of the process put it, scientists want to learn, "how the body mediates between the brain and the world." That understanding is expected to enhance their appreciation for how networks of neurons result in the ability of a living being—or our metal humanoid—to behave in a particular way. Along the way, such technologies will learn to learn from interacting with the physical world. In short, a mechanical intelligence exists; it developed along with the human brain and will be required to make our humanoid colleagues productive members of human society. Right now, this is so much out in the future that scientists are debating how to make the journey, and if it can be done.[20]

Creators of much of the early thinking about AI were wise enough to know the limits posed by information technologies and concepts, such as Julian Bigelow (1913–2003) and Norbert Wiener (1894–1964). George Dyson, a realistic observer of the IT scene, gets credit for rescuing from anonymity an observation made by Wiener in his last year of life, that "the world of the future will be an ever more demanding struggle against the limitations of our intelligence, not a comfortable hammock in which we can lie down to be waited upon by our robot slaves."[21] In 2020, Dyson added, "Those who seek to become minders over robots may end up minded by robots, instead."[22] Considering where AI and robotics are today, the long-term trend of information being consolidated, discussed in this book and that is one response to the fragmentation of specialized facts in the twentieth century, was now made possible by the massive aggregating capabilities of today's computers. Their capability lends credence to the comments of all three observers.

Meanwhile, as one science reporter concluded in 2024, "the humanoid revolution is upon us—expect these droids to start moving into industrial settings over the next few years, and then retail stores and our homes."[23] It seems to be a logical path as the technology improves, while giving all of us time to understand the new functions and their implications for society, jobs, and human existence.

CONCLUSIONS

Generative AI, when coupled with massive amounts of digitized data and, in the years to come, also to increasing amounts of computing power (called quantum computing), will continue to cause changes in what is done by humans. Some of their functions will continue to be delegated to myriad information technologies, while new activities will become possible. Vast amounts of new information will be created and studied, so the promise of more efficiency in matters such matters as environmentally responsible practices, the operation of machines (including vehicles), and medical practices should be expected. We come back to the long-standing aspiration expressed in the epigraph to this chapter: we live in a world in which there continues to be an increasing amount of information that needs to be useful and, most important, truthful. That is why the first six chapters of this book were necessary to understand what is happening now.

Notes

CHAPTER 1

1. Theodore Besterman et al. (eds.), *Oeuvres complètes de Voltaire,* 144 vols. (multiple publishers, 1968–2018).

2. Richard Van Noorden, "Global Scientific Output Doubles Every Nine Years," *newsblog,* May 7, 2014, http://blogs.nature.com/news/2014/05/global-scientific-output-doubles-every-nine-years.html (Accessed October 9, 2021).

3. Johan S.G. Chu and James A. Evans, "Slowed Canonical Progress in Large Fields of Science," *PNAS,* October 12, 2021, www.pnas.org (Accessed October 19, 2021).

4. Benjamin F. Jones, "The Burden of Knowledge and the "Death of the Renaissance Man': Is Innovation Getting Harder?," *NBER Working Paper Series,* 11360, May 2005, http://www.nber.org/papers/w11360 (Accessed July 9, 2017).

5. Richard Hanania, "We Need to Save Expertise From the Experts," *New York Times,* September 20, 2021.

6. By "organized information," I mean information (facts, data) as presented in some logical way, such as in a table or a narrative with context, and by topic, such as in a book that only provides biological or historical information.

7. Peter Burke, *Ignorance: A Global History* (New Haven, CT: Yale University Press, 2023): 35.

8. Ibid., 36.

9. Peter J. Bowler, *Evolution: The History of an Idea* (Berkeley: University of California Press, 2003); Robert Bud, *The Uses of Life: A History of Biotechnology* (Cambridge: Cambridge University Press, 1993); Kevin Davies, *Cracking the Genome: Inside the Race to Unlock Human DNA* (New York: Free Press, 2001); Stephen Jay Gould, *The Structure of Evolutionary Theory* (Cambridge, MA: Belknap Press of Harvard University Press, 2002); Lois M. Magner, *A History of the Life Sciences,* 3rd ed. (New York: Marcel Dekker, 2002); Jan Sapp, *Genesis: The Evolution of Biology* (New York: Oxford University Press, 2003); James Bridle, *Ways of Being:*

Animals, Plants, Machines: The Search for a Planetary Intelligence (New York: Farrar, Straus and Giroux, 2022).

10. Summarized and extended further in James W. Cortada, *All the Facts: A History of Information in the United States Since 1870* (New York: Oxford University Press, 2016): 1–9; Luciano Floridi, *Information: A Very Short Introduction* (New York: Oxford University Press, 2010); Toni Weller, *Information History—An Introduction: Exploring an Emerging Field* (New York: Neal-Schuman, 2008).

11. For an extensive literature discussion, see Cortada, *All the Facts*, 575–609.

12. For a useful discussion of its evolution, Ronald R. Kline, "Cybernetics, Management Science, and Technology Policy: The Emergence of "Information Technology' as a Keyword, 1948–1985," *Technology and Culture* 47, no. 3 (July 2006): 513–535; see also William Aspray, "The Scientific Conceptualization of Information: A Survey," *IEEE Annals of the History of Computing* 7 (1985): 117–140.

13. Useful here is Chris Wiggins and Matthew L. Jones, *How Data Happened: A History from the Age of Reason to the Age of Algorithms* (New York: W.W. Norton, 2023).

14. Angus Fletcher, *Storythinking: The New Science of Narrative Intelligence* (New York: Columbia University Press, 2023): 4–7, 102.

15. Rosalind Williams, "Crisis: The Emergence of Another Hazardous Concept," *Technology and Culture* 62, no. 2 (April 2021): 535.

16. While I have tried several times to provide comprehensive definitions, my best effort might be in Cortada, *All the Facts*, 1–22 and again in *Birth of Modern Facts: How the Information Revolution Transformed Academic Research, Governments, and Businesses* (Lanham, MD: Rowman & Littlefield): 1–26.

17. Philip Tetlow, *The Web's Awake: An Introduction to the Field of Web Science and the Concept of Web Life* (Hoboken, NJ: John Wiley & Sons/IEEE Press, 2007): 103.

18. Ibid.

19. Ibid.

20. Andrew Dillon, *Understanding Users: Designing Experience Through Layers of Meaning* (New York: Routledge, 2023): 115.

21. Weller, *Information History*.

22. The key finding in Cortada, *All the Facts*.

23. Thomas S. Kuhn, *The Structure of Scientific Revolutions* (Chicago, IL: University of Chicago Press, 1962, 1970, 1996, and 2012). The first two editions are the substantive ones on his thinking.

24. Tetlow, *The Web's Awake*, 13.

25. Ibid.

26. It is an old complaint of too much information expressed for centuries but that continues today: Harvey J. Graff, *Undisciplining Knowledge: Interdisciplinarity in the Twentieth Century* (Baltimore, MD: Johns Hopkins University Press, 2015); James Bridle, *New Dark Age: Technology and the End of the Future* (New York: Verso, 2018).

27. Lorraine Daston, *Rules: A Short History of What We Live By* (Princeton, NJ: Princeton University Press, 2022), especially pp. 212–237.

28. Paul M. Dover, *The Information Revolution in Early Modern Europe* (Cambridge: Cambridge University Press, 2021).

29. Thomas Suddendorf, Jonathan Redshaw, and Adam Bulley, *The Invention of Tomorrow: A Natural History of Foresight* (New York: Basic Books, 2022): 20.

30. Ibid., 174.

31. Alex Csiszar, *The Scientific Journal* (Chicago, IL: University of Chicago Press, 2018); Dennis Duncan, *Index, A History of the A Bookish Adventure from Medieval Manuscripts to the Digital Age* (New York: W.W. Norton, 2022).

32. Sapp, *Genesis*.

33. Tetlow, *The Web's Awake*, xiii.

34. Margaret J. Wheatley, *Leadership and the New Science: Discovering Order in a Chaotic World* (San Francisco, CA: Berrett-Koehler, 2006): 93–112.

35. In the 1930s, Russian psychologist Alexander R. Luria noted that people who were not exposed to scientific thinking viewed information as objects, reflections of what he called the "concrete," each fact related to others by how they interacted with each other. So a dog was related to a rabbit in that the former chased the latter, while someone exposed to scientific thinking (e.g., patterns, correlations" Alexander R. Luria, *Cognitive Development: Its Cultural and Social Foundations* (Cambridge, MA: Harvard University Press, 1976). For an early and thorough explanation of how techniques, processes, and technological developments by artisans and craftsmen came about and evolved, see Jacques Ellul, *The Technological Society* (New York: Vintage, 1964).

36. Liliane Hilaire-Pérez, "The Codification of Techniques: Between Bureaucracy and the Markets in Early Modern Europe from a Global Perspective," *Technology & Culture* 62, no. 2 (2021): 442–468; Matteo Vallerini (ed.), *The Structures of Practical Knowledge* (Dordrecht: Springer, 2017); Celina Fox, *The Arts of Industry in the Age of Enlightenment* (New Haven, CT: Yale University Press, 2009).

37. Benjamin M. Friedman, *Religion and the Rise of Capitalism* (New York: Alfred A. Knopf, 2021): 91–93.

38. Nathan L. Ensmenger, *The Computer Boys Take Over: Computers, Programmers, and the Politics of Technical Expertise* (Cambridge, MA: MIT Press, 2010): 27–136.

39. Hilaire-Pérez, "The Codification of Techniques: Between Bureaucracy and the Markets in Early Modern Europe from a Global Perspective," 451.

40. Cortada, *All the Facts*.

41. Cortada, *Birth of Modern Facts*, 249–276.

42. P. De Weerd, E. Smith, and P. Greenberg, "Effects of Selective Attention on Perceptual Filling-in," *Journal of Cognitive Neuroscience* 18, no. 3 (2006): 335–347.

43. For a long list of these with citations to their descriptions, see "List of cognitive biases," *Wikipedia*, https://en.wikipedia.org/wiki/List_of_cognitive_biases (Accessed June 24, 2021); for an introduction to these concepts, Rudiger F. Pohl, *Cognitive Illusions: Intriguing Phenomena in Thinking, Judgment and Memory* (London: Routledge, 2017).

44. Jonathan Rauch, *The Constitution of Knowledge: A Defense of Truth* (Washington, DC: Brookings Institution Press, 2021): 60.

45. Jason Epstein, *Book Business: Publishing Past, Present, and Future* (New York: W.W. Norton, 2001); Max Roser, "Books," *Our World in Data* (2013), https://ourworldindata. Org/books (Accessed June 28, 2021).

46. The literature on literacy and education is extensive, but for an introduction to the subject, see Edward E. Gordon and Elaine H. Gordon, *Literacy in America: Historic Journey and Contemporary Solutions* (Westport, CT: Praeger, 2003); Carl E. Kaestle, Helen Damon-Moore, Lawrence C. Stedman, Katherine Tinsley, and William Vance Trollinger, Jr., *Literacy in the United States: Readers and Reading since 1880* (New Haven, CT: Yale University Press, 1991); Cortada, *All the Facts*, 578–579.

47. U.S. Census Bureau.

48. Ibid.

49. U.S. Bureau of the Census, Historical Statistics.

50. For a brief, statistically loaded summary, see Benjamin M. Friedman, *Religion and the Rise of Capitalism* (New York: Knopf, 2021): 268–271.

51. Robert J. Gordon, *The Rise and Fall of American Growth: The U.S. Standard of Living Since the Civil War* (Princeton, NJ: Princeton University Press, 2017 ed.), 1.

52. Thomas Weiss, "U.S. Labor Force Estimates and Economic Growth, 1800–1860," in Robert E. Gallman and John Joseph Wallis (eds.), *American Economic Growth and Standards of Living Before the Civil War* (Chicago, IL: University of Chicago Press, 1992): 32.

53. U.S. Bureau of Economic Analysis.

54. Mitra Toossi, "A Century of Change: The U.S. Labor Force, 1950–2050," *Monthly Labor Review* 125, no. 5 (May 2002): 15–28.

55. Joseph Sunra Copeland, "U.S. Book Production," *McSweeney's Quarterly*, February 21, 2011, https://www.mcsweeneys.net/articles/us-book-production (Accessed June 28, 2021).

56. Daniel R. Headrick, *When Information Came of Age* (New York: Oxford University Press, 2002); Joel Mokyr, *A Culture of Growth: The Origins of the Modern Economy* (Princeton, NJ: Princeton University Press, 2018) and his earlier, *The Gifts of Athena: Historical Origins of the Knowledge Economy* (Princeton, NJ: Princeton University Press, 2002).

57. I have discussed these concepts more extensively in James W. Cortada, *Building Blocks of Modern Society: History, Information Ecosystems, and Infrastructures* (Lanham, MD: Rowman & Littlefield, 2021).

58. James Bridle, *Ways of Being: Animals, Plants, Machines: The Search for a Planetary Intelligence* (New York: Farrar, Straus and Giroux, 2022): 11.

59. Ryan Light and James Moody, "Introduction," in Ryan Light and James Moody (eds.), *The Oxford Handbook of Social Networks* (New York: Oxford University Press, 2020): 3.

60. Ryan Light and James Moody, "Network Basics: Points, Lines, and Positions," Ibid., 17.

61. Arthur Eddington, *The Nature of the Physical World* (London: J.M. Dant & Sons, 1935): 350.

62. Rauch, *The Constitution of Knowledge*, 65.

63. Michael Polanyi, "The Republic of Science: Its Political and Economic Theory," *Minerva* 1, no. 1 (September 1962): 54–73.

64. Claude Lévi-Strauss, *La Pensée sauvage* (Paris: Plon, 1962).

65. I provide a half-dozen diverse case studies to demonstrate its effectiveness as a methodological tool, Cortada, *Building Blocks of Society*.

66. Steve Sloman and Philip Fernbach, *The Knowledge Illusion: Why We Never Think Alone* (New York: Penguin, 2017).

67. Ibid.

68. Rauch, *The Constitution of Knowledge*, 88–92.

69. For a description, see U.S. Bureau of Labor Statistics, *Handbook of Occupational Groups and Families* (U.S. Government Printing Office, December 2018), and for actual descriptions of jobs, see the *Occupational Outlook Handbook*, https://www.bls.gov/ooh/ (Accessed June 30, 2021).

70. Emily S. Rueb, "Cursive Seemed to Go the Way of Quills and Parchment. Now It's Coming Back," *New York Times*, April 13, 2019, https://www.nytimes.com/2019/04/13/education/cursive-writing.html (Accessed June 30, 2021).

71. For example, Mary Poovey, *A History of the Modern Fact: Problems of Knowledge in the Sciences of Wealth and Society* (Chicago, IL: University of Chicago Press, 1998); Adrian Johns, *The Nature of the Book: Print and Knowledge in the Making* (Chicago, IL: University of Chicago Press, 1998); Donald Case, *Looking for Information: A Survey of Research on Information Seeking, Needs, and Behavior* (Bingley: Emerald Group Publishing, 2012): 285–324.

72. Dagmar Schäer and Simona Valeriani, "Technology Is Global: The Useful & Reliable Knowledge Debate," *Technology and Culture* 62, no. 2 (2021): 329.

73. Ibid., 331.

74. For example, John V. Pickstone, "Working Knowledge Before and After Circa 1800: Practices and Disciplines in the History of Science, Technology, and Medicine," *Isis* 98, no. 3 (2007): 489–516.

75. L.H. Gunderson, C.S. Holling, and S.S. Light, *Barriers and Bridges to the Renewal of Ecosystems and Institutions* (New York: Columbia University Press, 1995); C.S. Holling, "Resilience of Ecosystems: Local Surprise and Global Change," in W.C. Clark and R.E. Munn (eds.), *Sustainable Development of the Biosphere* (Cambridge: Cambridge University Press, 1986).

76. FN OUP 2.

77. Defended on their behalf, as for scientists in general, by Rauch, *The Constitution of Knowledge*.

78. Rosalind Williams, "Crisis: The Emergence of Another Hazardous Concept," *Technology and Culture* 62, no. 2 (2021): 542–543.

79. Ibid., 543.

80. John Stillwell, *Mathematics and Its History,* 3rd ed. (London: Springer-Verlag, 2010); Stephen M. Stigler, *The History of Statistics: The Measurement of Uncertainty Before 1900* (Cambridge, MA: Harvard University Press, 1986); also his, *Statistics on the Table: The History of Statistical Concepts and Methods* (Cambridge, MA: Harvard University Press, 2002); Anders Hald in three volumes, *A History of Mathematical Statistics From 1750 to 1930* (Hoboken, NJ: John Wiley & Sons, 1998)

and *History of Probability and Statistics and Their Application before 1750* (Hoboken, NJ: John Wiley & Sons, 1990), and *A History of Parametric Statistical Inference from Bernoulli to Fisher, 1713–1935* (Berlin: Springer-Verlag, 2007); Theodore M. Porter, *Trust in Numbers: The Pursuit of Objectivity in Science and Public Life* (Princeton, NJ: Princeton University Press, 1995) and *The Rise of Statistical Thinking, 1820–1900* (Princeton, NJ: Princeton University Press, 2020).

81. John M. Colaw and John K. Ellwood, *Teacher's Manual of School Arithmetic* (Richmond, VA: B.F. Johnson Publishing Co, 1908), 19.

82. Alexander R. Luria, *Cognitive Development: Its Cultural and Social Foundations* (Cambridge, MA: Harvard University Press, 1976): 99. Ask any author of a nonfiction book and they will probably agree that education—or writing a book—causes one to categorize, that is to say, organize, what they know about the topic being discussed.

83. Friedman, *Religion and the Rise of Capitalism,* 350.

84. Andreas Killen, *Nervous System: Brain Science in the Early Cold War* (New York: HarperCollins, 2023): 230.

85. Esther Landhuis, "Scientific Literature: Information Overload," *Nature*, July 20, 2016, https://www.nature.com/articles/nj7612-457a (Accessed June 28, 2021).

86. Z. D. Stephens et al., "Big Data: Astronomical or Genomical?," *PLOS Biology* thirteen (July 7, 2015), https://journals.plos.org/plosbiology/article?id=10.1371/journal.pbio.1002195 (Accessed June 28, 2021).

87. "Key Medline Indicators," National Library of Medicine, https://wayback.archive-it.org/org-350/20200416170941/https://www.nlm.nih.gov/bsd/bsd_key.html (Accessed June 28, 2021); for 2022, https://www.nlm.nih.gov/bsd/medline_pubmed_production_stats.html (Accessed October 31, 2023).

88. For an outstanding history, see Theodor M. Porter, *Genetics in the Madhouse: The Unknown History of Human Heredity* (Princeton, NJ: Princeton University Press, 2020).

89. Levin Clément et al., "A Data-Supported History of Bioinformatics Tools," Cornell University Digital Libraries, July 18, 2018, p. two, https://arxiv.org/abs/1807.06808 (Accessed June 28, 2021).

90. Floridi, *Information,* 86.

91. Timothy Lenoir, "Shaping Biomedicine as an Information Science," in Mary Ellen Bowden, Trudi Bellardo Hahn, and Robert V. Williams (eds.), *Science Information Systems,* 27.

92. STEM is a widely used term for science, technology, engineering, and mathematics as one mega-collection of disciplines and professions.

93. Jonathan Haidt, *The Righteous Mind: Why Good People Are Divided by Politics and Religion* (New York: Pantheon, 2012): 90.

94. Sloman and Fernbach, *The Knowledge Illusion,* 209–210 also for expanded discussion.

95. Floridi, *Information,* 70.

96. Wheatley, *Leadership and the New Science,* 158.

97. E. Coen, "The Storytelling Arms Race: Origin of Human Intelligence and the Scientific Mind," *Heredity* 123, no. 1 (2019): 67–78; on the role of emotion, R. Olson,

Don't Be Such a Scientist: Talking Substance in an Age of Style (Washington, DC: Island Press, 2009).

98. For a strong advocacy for storytelling in an age of Big Data and AI activism, see Fletcher, *Storythinking*.

99. Kristian H. Nielsen, "Histories of Science Communication," *Histories* 2 (2022): 334–340.

100. Peter Burke, *Ignorance: A Global History* (New Haven, CT: Yale University Press, 2023): 83. Even in the humanities and social sciences, inaccessibility exists, driven by the vocabulary (jargon-laden) and way of telling stories that are less comprehensible to people outside the fraternity of a discipline. For an example your author considers egregious, see a history paper by Stephanie Decker, "Introducing the Eventful Temporality of Historical Research into International Business," *Journal of World Business* 57, no. 6 (October 2022): 101380.

101. Hilaire-Pérez, "The Codification of Techniques: Between Bureaucracy and the Markets in Early Modern Europe from a Global Perspective," 456.

102. Eric Schatzberg, *Technology: Critical History of a Concept* (Chicago, IL: University of Chicago Press, 2018), 1.

103. Ibid.

104. On the day I wrote the first draft of this section, I received my weekly digital copy of *The Economist,* with this week's unsolicited tutorial, "Off the Charts: The Best of Our Data Journalism," *The Economist,* June 29, 2021.

105. Gordon, *The Rise and Fall of American Growth*, 460.

106. Mark Miodownik, *Stuff Matters: Exploring the Marvelous Materials that Shape Our Man-Made World* (Boston, MA: Mariner Books, 2013): 224.

107. Ibid.

108. Philip Tetlow, *The Web's Awake: An Introduction to the Field of Web Science and the Concept of Web Life* (Hoboken, NJ: John Wiley & Sons/IEEE Press, 2007): 8.

109. Ibid.

110. Many have commented on this possibility, often in outrageous or highly speculative, sensationalized language that I find uncomfortable and not useful. However, I too have cautiously approached the subject to a limited degree in James W. Cortada, *Living with Computers: The Digital World of Today and Tomorrow* (Chamm: Springer, 2020): 67–70, 99–100.

111. With apologies to those who study the very real circumstance of technological determinism, it is the idea, for example, that if you are a Microsoft user, you will probably continue to be so because the cost in time and money to convert to another collection of technologies may be too great. The same applies in our private lives: to replace our electric stove with a gas one may also be economically not worth it in exchange for perceived benefits.

112. For a recent example, David Gugerli, *Wie die Welt in den Computer kam: Zur Entstehung digitaler Wirklichkeit* (Frankfurt: Fischer Verlag, 2018); on the role of technological determinism, Schatzberg, *Technology*, 136–141.

113. For example, Jennifer A. Kingston, "Coming Soon: A Programmable Army of Humanoid Robots," *AXIOS*, March 14, 2024.

114. J. Bradford DeLong, *Slouching Toward Utopia: An Economic History of the Twentieth Century* (New York: Basic Books, 2022): 62.

CHAPTER 2

1. James W. Cortada, *Birth of Modern Facts: How the Information Revolution Transformed Academic Research, Governments, and Businesses* (Lanham, MD: Rowman & Littlefield, 2023); Harry Collins, *Are We All Scientific Experts Now?* (Cambridge: Polity, 2014).
2. Joel A. Elvery, "Changes in the Occupational Structure of the United States: 1860 to 2015," *Economic Commentary* Number 2019–09 (Cleveland, OH: Federal Reserve Bank of Cleveland, June 26, 2019): 1–8.
3. Ibid.
4. Ibid., 1.
5. David Autor, Frank Levy, and Richard J Murnane, "Upstairs Downstairs: Computers and Skills on Two Floors of a Large Bank," *Industrial Labor Review* 55, no. 3 (April 2002): 432–447.
6. Elvery, "Changes in the Occupational Structure of the United States: 1860 to 2015," 2.
7. Ibid. Economist Robert J. Gordon also documented a slowdown in innovation after 1970 as well, in *Rise and Fall of American Growth: The U.S. Standard of Living since the Civil War* (Princeton, NJ: Princeton University Press, 2016).
8. For the details on how computing diffused around the world by country and region, see James W. Cortada, *The Digital Flood: The Diffusion of Information Technology Across the U.S., Europe, and Asia* (New York: Oxford University Press, 2012). It includes a detailed bibliographic essay about developments on a country-by-country basis, Ibid., 733–768.
9. Government agencies were the first adopters of computers, largely starting during World War II, but in small numbers. They did, however, fund the lion's share of R&D on computing in the 1930s–1950s.
10. There is a large literature on this topic; however, useful overviews include Martin Campbell-Kelly, William Aspray, Nathan Ensmenger, and Jeffrey R. Yost, *Computer: A History of the Information Machine,* 3rd ed. (Boulder, CO: Westview, 2014); Paul E. Ceruzzi, *A History of Modern Computing,* 2nd ed. (Cambridge, MA: MIT Press, 2003).
11. Martin Campbell-Kelly and Daniel D. Garcia-Swartz, *From Mainframes to Smartphones: A History of the International Computer Industry* (Cambridge, MA: Harvard University Press, 2015): 11–54.
12. I have commented extensively on this pre-computer time and on the early era of computing, James W. Cortada, *Before the Computer: IBM, NCR, Burroughs, and Remington Rand and the Industry They Created, 1865–1956* (Princeton, NJ: Princeton University Press, 1993): 44–63, 128-136222–246; *The Computer in the United States: From Laboratory to Market, 1930–1960* (Armonk, NY: M.E. Sharpe, 1993): 64–124; *Information Technology as Business History: Issues in the History and*

Management of Computers (Westport, CT: Greenwood Press, 1996): 141–200; *IBM: The Rise and Fall and Reinvention of a Global Icon* (Cambridge, MA: MIT Press, 2019): 177–202.

13. Cortada, *IBM*, 194, 225–226, 286–287.

14. Cortada, *Information Technology as Business History*. As part of my responsibilities at IBM in the 1970s and 1980s, I observed, and later wrote about, how data processing organizations were run and operated. For what I learned, James W. Cortada, *EDP Costs and Charges: Finance, Budgets, and Cost Control in Data Processing* (Englewood Cliffs, NJ: Prentice-Hall, 1980), *Managing DP Hardware: Capacity Planning, Cost Justification, Availability, and Energy Management* (Englewood Cliffs, NJ: Prentice-Hall, 1983), *Strategic Data Processing: Considerations for Management* (Englewood Cliffs, NJ: Prentice-Hall, 1984): 53–86.

15. General Electric, *The Next Step in Management: An Appraisal of Cybernetics* (General Electric, 1952, 1955): 91.

16. Reducing labor content of work through automation continues to the present, as described in Chapters 6 and 7 with the use of robotics and AI.

17. J. Bénay, "General Considerations on the Economics of A.D.P.," in A.B. Frielink (ed.), *Economics of Automated Data Processing* (Amsterdam: North-Holland Publishing Co., 1965): 10–21; Ned Chapin, "Justifying the Use of an Automatic Computer," *Journal of Machine Accounting, Systems and Management* 6, no. 8 (September 1955): 9–10, 14; Robert J. Gordon, "The Postwar Evolution of Computer Prices," in Dale W. Jorgenson and Ralph Landau (eds.), *Technology and Capital Formation* (Cambridge, MA: MIT Press, 1989): 77–125; Frank M. Knox, "The Integrated Approach to Office Cost Control," in *The Changing Dimensions of Office Management,* AMA Management Report Number 41 (New York: Office Management Division, American Management Association, 1960): 129–134; Harold Leavitt and Thomas J. Whisler, "Management in the 1980s," *Harvard Business Review* 36, no. 6 (November-December, 1958): 41–48, probably the most widely read article on the subject in the early years of computing; Linda Runyan, "40 Years on the Frontier," *Datamation* 37, no. 6 (March 15, 1991): 34–57; Douglas J. Axsmith, "Economic Aspects of Data Processing," in *Data Processing: Proceedings 1963* (Detroit: Data Processing Management Association, 1963): 70–91; Rudolph E. Hirsch, "The Value of Information," *Journal of Accountancy* 125 (June 1968): 41–45.

18. Roddy F. Osborn, "GE and UNIVAC: Harnessing the High-Speed Computer," *Harvard Business Review* 32, no. 4 (July–August 1954): 99–107.

19. I explore all of these issues in James W. Cortada, *The Digital Hand: How Computers Changed the Work of American Manufacturing, Transportation, and Retail Industries* (New York: Oxford University Press, 2004) and in vol. 2 *The Digital Hand: How Computers Changed the Work of American Financial, Telecommunications, Media, and Entertainment Industries* (New York: Oxford University Press, 2006); Erik Brynjolfsson and Shinkyu Tang, "Information Technology and Productivity: A Review of the Literature," in Marvin Zelkwitz (ed.), *Advances in Computers* 43 (Cambridge, MA: Academic Press, 1996).

20. The literature is vast, but for a sense of activity, see Robert V. Head, "Old Myths and New Realities in Business Applications," *Datamation* 13, no. 9

(September 1967): 26–28; Leonard Rico, *The Advance Against Paperwork: Computers, Systems, and Personnel* (Ann Arbor: Graduate School of Business Administration, University of Michigan, 1967); R.A. Hirschheim, *Office Automation: A Social and Organizational Perspective* (New York: John Wiley & Sons, 1985); Cortada, *Information Technology as Business History,* 141–187.

21. J.F. Rockart and D.W. DeLong, *Executive Support Systems: The Emergence of Top Management Computer Use* (Homewood, IL: Dow Jones-Irwin, 1988); P.G.W. Keen and M.S. Morton, *Decision Support Systems: An Organizational Perspective* (Reading, MA: Addison-Wesley, 1978).

22. Cortada, *Digital Hand,* vol. 2, 37–150.

23. Ibid.

24. For a study of one part of the insurance industry, JoAnne Yates, *Structuring the Information Age: Life Insurance and Technology in the Twentieth Century* (Baltimore, MD: Johns Hopkins University Press, 2008); Cortada, *Digital Hand,* vol. 2, 13–150.

25. Data entry awaits its historian, who will find much contemporary commentary on the subject. For a bibliography of this material, see James W. Cortada, *Second Bibliographic Guide to the History of Computing, Computers, and the Information Processing Industry* (Westport, CT: Greenwood Press, 1996): 302–305.

26. Cortada, *The Digital Hand.*

27. C.T. Carter, "What's Ahead for Liquids Pipe Line Automation," *Pipe Line Industry* Part 1 (April 1973): 25–27, Part II (May 1973): 69–72; Meyer Glenn, "SCADA: A 10 Year Look into the Future," Ibid. (April 1985): 19–23; E.M. Smith and D.T. Sweeney, "Computer Helps Design, Run Pipeline," *Oil and Gas Journal* (December 8, 1975): 94–106.

28. C.E. Bodington and T.E. Baker, "A History of Mathematical Programming in the Petroleum Industry," *Interfaces* 20, no. 4 (July–August 1990): 117–127; Thomas M. Stout, "Process Control: Past, Present, and Future," *The Annals of the American Academy of Political and Social Science* 340 (March 1962): 29–37; William E. Miller (ed.), *Digital Computer Applications to Process Control* (New York: Instrument Society of America, 1965).

29. For an example from an early survey, T. M. Whittin, "Report on an Inventory Management Survey," *Production and Inventory Management* 7, no. 1 (January 1966): 27–32; for a bibliography on early applications, James W. Cortada, *A Bibliographic Guide to the History of Computer Applications, 1950–1990* (Westport, CT: Greenwood Press, 1996): 177–179.

30. I have commented more fully on the broad sweep of this important story in *IBM,* 203–231.

31. Ralph Watson McElvenny and Marc Wortman, *The Greatest Capitalist Who Ever Lived: Tom Watson Jr. and the Epic Story of How IBM Created the Digital Age* (New York: PublicAffairs, 2023): 339–394.

32. On the technical capabilities of the System 360 family of computers, Emerson W. Pugh, Lyle R. Johnson, and John H. Palmer, *IBM's 360 and Early 370 Systems* (Cambridge, MA: MIT Press, 1991).

33. Cortada, *Information Technology as Business History.* New lines of service businesses developed rapidly too, Jeffery R. Yost, *Making IT Work: A History of the Computer Services Industry* (Cambridge, MA: MIT Press, 2017); Charles P. Bourne and Trudi Bellardo Hahn, *A History of Online Information Services, 1963–1976* (Cambridge, MA: MIT Press, 2003).

34. Franklin M. Fisher, James W. McKie, and Richard B. Mancke, *IBM and the U.S. Data Processing Industry: An Economic History* (New York: Praeger, 1983): 143–361.

35. This was a worldwide phenomenon I found in every industrializing society when exploring how computing diffused around the world, Cortada, *The Digital Flood.*

36. This entire section is based on my study of numerous Bureau of Labor Statistics reports on these various industries that I summarize in Cortada, *Information Technology as Business History,* 170–176. I also reported their findings in each of the more than one dozen industries I examined for the 3 volume, *The Digital Hand.*

37. Cortada, *Information Technology as Business History,* 170–176.

38. These were the activities I described in the three volumes of Cortada, *The Digital Hand.* For a bibliography on these applications, see Cortada, *A Bibliographic Guide to the History of Computer Applications, 1950–1990.* The latter contains 1,649 citations and was not definitive, merely representative of the amount of commentary appearing on the use of computers in the second half of the twentieth century.

39. Cortada, *Digital Hand,* 105, 110–111, 115–117.

40. The implications of such changes were not lost on observers of the role of technology and management. See James M. Utterback, *Mastering the Dynamics of Innovation* (Boston, MA: Harvard Business School Press, 1994); Clayton M. Christensen, *The Innovator's Dilemma: When New Technologies Cause Great Firms to Fail* (Boston, MA: Harvard Business School Press, 1997); Carl Shapiro and Hal R. Varian, *Information Rules: A Strategic Guide to the Network Economy* (Boston, MA: Harvard Business School Press, 1999); Thomas S. Wurster, *Blown to Bits: How the New Economics of Information Transforms Strategy* (Boston, MA: Harvard Business School Press, 2000). More tactically about how computers were being used, see James W. Cortada, *Best Practices in Information Technology: How Corporations Get the Most Value from Exploiting Their Digital Investments* (Upper Saddle River, NJ: Prentice Hall PTR, 1998) and *21st Century Business: Managing and Working in the New Digital Economy* (Upper Saddle River, NJ: Financial Times/Prentice Hall, 2001).

41. For the technology side of the story, Ceruzzi, *A History of Modern Computing,* 207–241; for the business volumes aspect, Campbell-Kelly and Swartz, *From Mainframes to Smartphones;* for the IBM experience, *IBM,* 379–418.

42. For example, in 1984–1985 your author was a sales manager for IBM in Nashville, Tennessee, leading a team of salesmen focused on manufacturing accounts. We sold 6,000 PCs to American Standard Corporation, a manufacturer of bathroom fixtures, for installation in all their dealers in the United States. We literally had conversations with the IBM PC factory in Boca Raton, Florida, asking how many PCs could be shipped in an 18-wheel truck. In the second half of the decade, not only

could one see the newest models being delivered in bulk to an office building, but one could also view older models stacked up in closets and storerooms waiting to be shipped out to recycling companies or to municipal trash dumps.

43. Cortada, *The Digital Flood*, 496, 500, 511, 517, 527, 550.

44. Thomas Haigh, "A Veritable Bucket of Facts: Origins of the Data Base Management System, 1960–1980," in W. Boyd Rayward and Mary Ellen Bowden (eds.), *The History and Heritage of Scientific and Technological Information Systems* (New Medford, NJ: Information Today, 2004): 73–88; Thomas J. Bergin and Thomas Haigh, "The Commercialization of Database Systems, 1969–1983," *IEEE Annals of the History of Computing* 31, no. 4 (October–December 2009): 26–41; on technical issues, any edition of C.J. Date, *An Introduction to Database Systems,* here I consulted 6th ed. (Reading, MA: Addison-Wesley, 1995).

45. Stephen A. Brown, *Revolution at the Checkout Counter: The Explosion of the Bar Code* (Cambridge, MA: Harvard University Press, 1997).

46. Definitions became less meaningful. One zettabyte is equal to 1,000 exabytes or to 1 billion terabytes. One terabyte is equal to one trillion bytes, while one byte is equal to 8 bits (the same as one character, such as a, b, c, 1, 2, 3).

47. But for an example of this kind of literature, see "The Big Data Facts Update 2020," https://www.nodegraph.se/big-data-facts/ (Accessed May 6, 2021).

48. The influential texts of the day included Thomas H. Davenport, *Process Innovation: Reengineering Work through Information Technology* (Boston, MA: Harvard Business Review Press, 1992); Michael Hammer, *Beyond Reengineering: How the Process-Centered Organization is Changing Our Work and Our Lives* (New York: Harper Business, 1996).

49. The volume of publications on these events was voluminous, which I tracked with a colleague for several years, for example, John A. Woods and James W. Cortada (eds.), *The 1998 ASTD Training and Performance Yearbook* (New York: McGraw-Hill, 1998); we continued this yearbook through 2002.

50. Cortada, *Information and the Modern Corporation*, 33–54.

51. David Blanchard, *Supply Chain Management Best Practices* (Hoboken, NJ: John Wiley & Sons, 2010); Elaine M. Lai, "An Analysis of the Department of Defense Supply Chain: Potential Applications of the Auto-ID Center Technology to Improve Effectiveness," Unpublished B.A. thesis, MIT, 2003, pp. 17–36.

52. Jeffrey Liker, *The Toyota Way: 14 Management Principles from the World's Great Manufacturer* (New York: McGraw-Hill, 2004).

53. Discussions about post-pandemic supply chains began immediately with the medical crisis because they were so pervasive and ubiquitous. See, for example, Esther Shein, "How U.S. Supply Chains Can Improve Operations Post-Pandemic," *TechRepublic*, April 21, 2020, https://www.techrepublic.com/article/how-us-supply-chains-can-improve-operations-post-pandemic/ (Accessed May 8, 2021); Robert J. Bowman, "How Will Supply Chains Adjust to the Post-Pandemic Age?" *SupplyChainBrain*, May 6, 2020, https://www.supplychainbrain.com/articles/31257-how-will-supply-chains-adjust-to-the-post-pandemic-age (Accessed May 8, 2021).

54. On its early history, Henry Parker Willis, *The Federal Reserve System: Legislation, Organization and Operation* (New York: Ronald Press, 1923): 598–824;

on how it works today, Federal Reserve System, *The Federal Reserve System Purposes & Functions,* 20th ed. (Washington, DC: U.S. Federal Reserve System, 2019): 118–151. The first edition of this publication appeared in May 1939.

55. I explain his ideas in James W. Cortada, *Birth of Modern Facts: How the Information Revolution Transformed Academic Research, Governments, and Businesses* (Lanham, MD: Rowman & Littlefield, 2023), 106–112.

56. Cortada, *The Digital Hand,* vol. 1.

57. For example, in the spring of 2020, as the Covic-19 pandemic spread in the United States, stores ran out of toilet paper, meat shortages appeared, and it was impossible to obtain adequate medical supplies and masks, yet that had not been a problem before. Toilet paper is a good example. Manufacturers had known for decades exactly how much was needed and when everywhere and so production around the world had been finely tuned to meet the accurately known demand for this product in a reliable way for decades. When people started hoarding it, the supply chains around the world could not respond quickly, yet people continued to use toilet paper at pre-pandemic rates of consumption. Brent Schrotenboer, "Coronavirus and Shopping for Supplies: Getting to the Bottom of the Toilet Paper Shortage," *U.S.A. Today,* April 8, 2020, https://www.usatoday.com/story/money/2020/04/08/coronavirus-shortage-where-has-all-the-toilet-paper-gone/2964143001/ (Accessed May 6, 2021).

58. Michael Hugos, *Essentials of Supply Chain Management,* 2nd ed. (Hoboken, NJ: John Wiley & Sons, 2006): 133–168; Jack R. Meredith and Scott M. Shafer, *Operations and Supply Chain Management for MBAs,* 7th ed. (Hoboken, NJ: John Wiley & Sons, 2019): 240–328.

59. Victor Zarnowitz, *Business Cycles: Theory, History, Indicators, and Forecasting* (Chicago, IL: University of Chicago Press, 1996): 183–202 for the concepts; Christopher M. Bishop, *Pattern Recognition and Machine Learning* (Berlin: Springer-Verlag, 2006) for a technical "how to" text; the subject awaits its historian.

60. I discussed the role of librarians in considerable detail in Cortada, Birth of Modern Facts, 57–87.

61. Karen Butner and Dave Lubowie, *Welcome to the Cognitive Supply Chain: Digital Operations—Reimagined* (Armonk, NY: IBM Corporation, 2017).

62. The core of Cortada, *Information and the Modern Corporation,* and a sequel study, James W. Cortada, *The Essential Manager: How to Thrive in the Global Information Jungle* (Hoboken, NJ: John Wiley & Sons, 2015): 40–123.

63. This circumstance is analogous to bathroom fixtures travelers encounter in other countries. They know what the fixtures are used for, but have to learn which knobs to turn and how to do that; then they are back in business as if in their home country.

64. On the earlier number, *flexjobs,* https://www.flexjobs.com/blog/post/remote-work-statistics/; on the second, "Gartner HR Reveals 41% of Employees Likely to Work Remotely at Least Some of the Time Post Coronavirus Pandemic," Press Release, Gartner, April 14, 2020, https://www.gartner.com/en/newsroom/press-releases/2020-04-14-gartner-hr-survey-reveals-41--of-employees-likely-to-; Richard Eisenberg, "Is Working From Home The Future of Work?," *Forbes,* April 10,

2021, https://www.forbes.com/sites/nextavenue/2020/04/10/is-working-from-home-the-future-of-work/#3ab6721b46b1 (All accessed May 9, 2021).

65. For a brief analysis of its origins, see Stephen Lukasik, "Why the Arpanet was Built," *IEEE Annals of the History of Computing* 33, no. 3 (July–September 2011): 4–21; on its commercialization, refer to Shane Greenstein, *How the Internet Became Commercial: Innovation, Privatization, and the Birth of a New Network* (Princeton, NJ: Princeton University Press, 2015).

66. The increase in expenditures progressed at various rates in different industries and times, a key finding of Cortada, *The Digital Hand,* 3 vols.

67. Germaine Halegoua, *Smart Cities* (Cambridge, MA: MIT Press, 2020): 85–124; Ben Green, *The Smart Enough City: Putting Technology in Its Place to Reclaim Our Urban Future* (Cambridge, MA: MIT Press, 2019): 1–14, 143–164; and for a series of interviews with experts and public officials, Stefano L. Tresca, *Future Cities: 42 Insights and Interviews with Influencers, Startups, Investors* (New York: Seahorse Press, 2017).

68. Cortada, *Information and the Modern Corporation*, 80–84. The conversation had been underway for many years, James Arlin Cooper, *Computer and Communications Security: Strategies for the 1990s* (New York: McGraw-Hill, 1989); Thomas W. Madron, *Network Security in the 90s: Issues and Solutions for Managers* (Hoboken, NJ: John Wiley & Sons, 1992).

69. Digital humor has yet to be methodically documented, but see Chris Miksanek, *400 Years of Computer Humor* (No city: Self Published, 2008, 2012); Marilyn K. Martin and Jack Dunning, *The Best Computer Humor On The Web!* (San Diego, CA: ComputerEdge E-Books, 2014); Teresa Correa, "Bottom-Up Technology Transmission Within Families: Exploring How Youths Influence Their Parents' Digital Media Use With Dyadic Data," *Journal of Communication* 64, no. 1 (February 2014): 103–124 and that has an extensive bibliography. For one of the earliest important studies, see John Palfrey and Urs Gasser, *Born Digital: Understanding the First Generation of Digital Natives* (New York: Basic Books, 2008, 2016).

70. Humor on all aspects of computing has a rich history yet to be properly studied. While at IBM for nearly 40 years, I quietly collected paper cartoons and other ephemera floating around IBM and customer offices, with the result that I now have over 1,000 items.

71. For those not familiar with the phrase *coalface,* it is about doing the work in a job where it actually takes place rather than discussing or planning for it. If you were going to solve hunger or supply coal miners with food and medical care, you went to the towns where coal miners lived and worked, talking and working with these individuals whose faces were black from coal dust just acquired in the mines, hence the term *coalface*. Americans use the phrase "where the action is." Both call for managers and others "being in touch" with the day-to-day processes of an enterprise.

72. IBM's CEO in the early 2000s famously made the point that no CEO could order or direct employees, because there were too many of them and they knew more than the senior executive, and so had to be persuaded to move in desired directions. Samuel J. Palmisano, "The Globally Integrated Enterprise," *Foreign Affairs* 84, no. 3 (May/June 2006).

73. There are over 8 billion people, but there are over 30 billion sensors and other devices connected to the Internet as of 2020, often understood as the "Internet of things" (IoT), "Internet of Things (IoT) Connected Devices Installed Base Worldwide," *Statista,* https://www.statista.com/statistics/471264/iot-number-of-connected-devices-worldwide/ Accessed May 9, 2021).

74. Paul N. Edwards, *The Closed World: Computers and the Politics of Discourse in Cold War America* (Cambridge, MA: MIT Press, 1996); Colin B. Burke, *America's Information Wars: The Untold Story of Information Systems in America's Conflicts and Politics from World War II to the Internet Age* (Lanham, MD: Rowman & Littlefield, 2018), 117–136, 277–336.

75. Brian McCullough, *How the Internet Happened: From Netscape to the iPhone* (New York: W.W. Norton, 2018), 7–37; Katie Hafner and Matthew Lyon, *Where Wizards Stay Up Late: The Origins of the Internet* (New York: Simon & Schuster, 1996); Janet Abbate, *Inventing the Internet* (Cambridge, MA: MIT Press, 1999), 7–42, 113–146.

76. Shane, *How the Internet Became Commercial,* Cortada, *All the Facts,* 415–454; Barry Wellman and Caroline Haythornthwaite (eds.), *The Internet in Everyday Life* (Oxford: Blackwell, 2002); Lee Rainie and Barry Wellman, *Networked: The New Social Operating System* (Cambridge, MA: MIT Press, 2012); William Aspray and Barbara M. Hayes (eds.), *Everyday Information: The Evolution of Information Seeking in America* (Cambridge, MA: MIT Press, 2011).

77. For example, the Pew Research Center (scholarly and rigorous), Internet World Stats (marketing services), and International Telecommunications Union (industry organization).

78. My research on the global diffusion of computing uncovered that almost no country on Earth remained a light user of smartphones. The poorer the country, the more people went straight to smartphones, bypassing the phase of first accessing the Internet through PCs (Cortada, *The Digital Flood).* In 2020, there were an estimated 3.5 billion smartphones in use worldwide, amounting to 45 percent of all humans owning one, which, if we subtract our little children, essentially takes that percentage to over 60, "Number of Smartphone Users Worldwide from 2016 to 2021," *Statista,* https://www.statista.com/statistics/330695/number-of-smartphone-users-worldwide/ Accessed May 9, 2021).

79. Rita Gunther McGrath, "The Pace of Technology Adoption is Speeding Up," *Harvard Business Review,* November 25, 2013, https://hbr.org/2013/11/the-pace-of-technology-adoption-is-speeding-up (Accessed September 1, 2020); Angela Hausman, "Innovation: Adoption and Diffusion in the Age of Social Media," *Market Maven,* (undated), https://www.hausmanmarketingletter.com/innovation-adoption-diffusion-age-social-media/ (Accessed September 1, 2020).

80. Hee-Woong Kim, Sumeet Gupta, and Joon Koh, "Investigating the Intention to Purchase Digital Items in Social Networking Communities: A Customer Value Perspective," *Information & Management* 48 (2011): 228–234; Ofir Turel, Alexander Serenko, and Nick Bontis, "User Acceptance of Hedonic Digital Artifacts: A Theory of Consumption Values Perspective," Ibid., 47 (2010), 53–59; R. Agarwal and E. Karahanna, "Time Flies When You Are Having Fun: Cognitive Absorption

and Beliefs About Information Technology Usage," *MIS Quarterly* 24, no. 4 (2000): 665–694; V. Venkatesh and S.A. Brown, "A Longitudinal Investigation of Personal Computers in Homes: Adoption Determinants and Emerging Challenges," Ibid., 25, no. 1 (2001): 71–102; C. Page Moreau, Donald R. Lehmann, and Arthur B. Markman, "Entrenched Knowledge Structures and Consumer Response to New Products," *Journal of Marketing Research* 38 (February 2001): 14–29; Jiewen Hong and Brian Sternthal, "The Effects of Consumer Prior Knowledge and Processing Strategies on Judgments," *Journal of Marketing Research* 47, no. 2 (April 2010): 301–311; Steve Hoeffler, "Measuring Preferences for Really New Products," Ibid., 40, no. 4 (November 2003): 406–420.

81. Data from the International Data Corporation, one of the most widely consulted sources for such information, https://www.idc.com/about (Accessed May 9, 2020).

82. Another reliable source, International Telecommunications Union, https://www.itu.int/en/about/Pages/default.aspx (Accessed May 9, 2021).

83. *Microsoft 2016 Annual Report* (Seattle, WA: Microsoft Corporation, 2017), unpaginated.

84. Mark Muro, Sifan Liu, Jacob Whiton, and Siddharth Kulkarni, *Digitization and the American Workforce* (Washington, DC: Brookings Institution, November 2017): 6.

85. Ibid.

86. Ibid.

87. The number of transistors on a microchip doubles essentially every two years, while their cost drops by half in the same period. Put another way, one can expect the speed and function of a computer to increase substantially every two years, while the cost for such additional capability drops at predictable rates, such as by half.

88. This is a subject of my current research, documenting this consumer behavior. For my initial reporting, see James W. Cortada, "Can Moore's Law Teach Us How Users Decide When to Acquire Digital Devices?," *Interfaces* four (2023), https://cse.umn.edu/cbi/interfaces#Moore (Accessed November 1, 2023).

89. Eliza Paul, "What is Digital Signature—How it Works, Benefits, Objectives, Concept," *EMP Trust HR,* September 12, 2017, https://www.emptrust.com/blog/benefits-of-using-digital-signatures (Accessed May 9, 2021).

90. Also, for a discussion of standards, an age-old issue in information's transformation in modern times, see Dawn M. Turner, "Major Standards and Compliance of Digital Signatures—A World-wide Consideration," *Cryptomathic,* January 7, 2016, https://www.cryptomathic.com/news-events/blog/major-standards-and-compliance-of-digital-signatures-a-world-wide-consideration (Accessed May 9, 2021).

91. For the initial text of the law, U.S. Government Publishing Office, "Public Law 106-229—June 30, 2000," entitled "Electronic Signatures in Global and National Commerce Act," https://www.govinfo.gov/content/pkg/PLAW-106publ229/pdf/PLAW-106publ229.pdf (Accessed May 9, 2021).

92. Muro, Liu, and Kulkarni, *Digitization and the American Workforce,* 10–14.

93. Ibid., 12.

94. U.S. Bureau of Labor Statistics, *Occupational Outlook Handbook*, published in many editions since the 1870s and in recent decades online, https://www.bls.gov/ooh/ (Accessed May 9, 2021).

95. Muro, Liu, and Kulkarni, *Digitization and the American Workforce*, 15.
96. Ibid., 17.
97. Ibid., 21.
98. James K. Willcox, "Libraries and Schools Are Bridging the Digital Divide During the Coronavirus Pandemic," *CR Consumer Reports,* April 29, 2020, https://www.consumerreports.org/technology-telecommunications/libraries-and-schools-bridging-the-digital-divide-during-the-coronavirus-pandemic/; Laura Isensee, "Parents, Educators Worry Digital Divide During COVID-19 Pandemic Will Widen Learning Gaps," *Houston Public Media,* April 27, 2020, https://www.houstonpublicmedia.org/articles/education/2020/04/27/367491/parents-educators-worry-digital-divide-during-the-pandemic-will-widen-learning-gaps/; Maeve Reston, "Pandemic Underscores Digital Divide Facing Students and Educators," *CNN,* April 9, 2020, https://www.cnn.com/2020/04/09/politics/digital-divide-education-coronavirus/index.html (All Accessed May 9, 2021).

99. For example, Internet Live Stats, https://www.internetlivestats.com/google-search-statistics/ (Accessed May 11, 2021).

100. Some became cultural icons, such as those produced by the late Hans Rosling 1948–2017), that must be seen. Available on YouTube, https://www.youtube.com/watch?v=hVimVzgtD6w, https://www.youtube.com/watch?v=jbkSRLYSojo, https://commons.wikimedia.org/wiki/Category:Hans_Rosling (Accessed May 12, 2021).

101. I have commented on this style more extensively in Cortada, *The Essential Manager*, especially on pages 1–84.

102. Collins, *Are We All Scientific Experts Now?*, 45. A personal story: a colleague of mine who was respected by a university press editor wrote to him to say that I was the leading expert about IBM Corporation "on the planet" and that he should publish my work. The next day, that editor reached out to me even though we had never met and he had never read any of my books.

CHAPTER 3

1. Arnold Pacey, *The Culture of Technology* (Cambridge, MA: MIT Press, 1983): 162.

2. Michael Buckland, *Information and Society* (Cambridge, MA: MIT Press, 2017): xiii.

3. I commented more fully on the human-computer relationship and its future implications in James W. Cortada, *Living with Computers: The Digital World of Today and Tomorrow* (Cham: Springer, 2020).

4. Ella Koeze and Nathan Popper, "The Virus Changed the Way We Internet," *New York Times*, April 7, 2020, https://www.nytimes.com/interactive/2020/04/07/technology/coronavirus-internet-use.html (Accessed May 20, 2021).

5. James W. Cortada, *How Societies Embrace Information Technology: Lessons for Management and the Rest of Us* (Hoboken, NJ: John Wiley & Sons, 2009): 165–167.

6. "Pat" (*Saturday Night Live*), *Wikipedia* https://en.wikipedia.org/wiki/Pat_(Saturday_Night_Live) (Accessed May 15, 2021).

7. The origins of the renaming of the world wars may have been started by historian Eric Hobsbawm, *The Age of Extremes: The Short Twentieth Century, 1914–1991* (London: Michael Joseph, 1994); but see also Anthony Shaw and Ian Westwell, *The World in Conflict, 1914–1945* (London: Routledge, 2000); P.M.H. Bell, *The Origins of the Second World War in Europe*, 3rd ed. (London: Routledge, 2015): 15–42; Philip Bobbitt, *The Shield of Achilles: War, Peace, and the Course of History* (New York: Knopf, 2002): 34–44.

8. For a gem of an example by a highly experienced historian, see Gerda Lerner, *Why History Matters: Life and Thought* (New York: Oxford University Press, 1997).

9. See how a half dozen experienced historians tiptoed around the issue by looking at the title of their book, Alfred D. Chandler, Jr. and James W. Cortada (eds.), *A Nation Transformed by Information: How Information Has Shaped the United States from Colonial Times to the Present* (New York: Oxford University Press, 2000).

10. Tom Standage, *The Victorian Internet: The Remarkable Story of the Telegraph and the Nineteenth Century's On-Line Pioneers* (New York: Berkley Books, 1998).

11. Historians have written about naming issues, but for a recent discussion by a highly regarded one, see the comments of the late Arthur Marwick in *The Sixties: Cultural Revolution in Britain, France, Italy, and the United States, c. 1958-c. 1974* (Oxford: Oxford University Press, 1998), in which he walks the reader through the issues he had to consider when declaring the 1960s a distinct period, pp. 3–38. I have been highly influenced by his thinking on the matter.

12. The subject of Marwick's book, *The Sixties*.

13. Buckland, *Information and Society*, 16.

14. Thereby fundamentally changing the shape of their bodies in subsequent millennia by reducing the size of the digestive tract, making more energy available and enabling the brain to grow in size. Richard Wrangham, *Catching Fire: How Cooking Made Us Human* (New York: Basic Books, 2009): 105–128.

15. Ibid., 169.

16. For example, Buckland, *Information and Society*, which I draw from for this exercise.

17. The core argument made recently by two distinguished economists, Joseph E. Stiglitz and Bruce C. Greenwald, in *Creating a Learning Society: A New Approach to Growth, Development, and Social Progress* (New York: Columbia University Press, 2014): 4–5.

18. Tracked by the International Telecommunication Union.

19. Statista, https://www.statista.com/statistics/264810/number-of-monthly-active-facebook-users-worldwide/ (Accessed May 15, 2021).

20. Stiglitz and Greenwald, *Creating a Learning Society*, 5–6. The italics are from the original quote.

21. For an extensive bibliography of their publications and of others, see Ibid., 587–623.

22. Ibid., 26.

23. I collected evidence of these trends for the United States in James W. Cortada, *All the Facts: A History of Information in the United States Since 1870* (New York: Oxford University Press, 2016).

24. For an anthology of seminal literature on this topic, see James W. Cortada (ed.), *Rise of the Knowledge Worker* (Boston, MA: Butterworth-Heinemann, 1998).

25. They were coming into my home at the rate of a half dozen per week for so long that I donated the collection to the Hagley Museum & Library. Since making this donation in 2018, over 500 more came into my library by early 2020. For details, see https://www.hagley.org/librarynews/new-collection-james-w-cortada-collection-information-technology-publications (Accessed May 15, 2021). On a personal note, I began collecting books discussing increased uses of information in the 1970s, and by 2010 I had accumulated over 4,000 volumes from all corners of the Earth, contributing to my own information load, while changing how I conducted research and what issues I needed to study.

26. One of the best overviews of the subject is by Frank Webster, *Theories of the Information Society,* 4th ed. (London: Routledge, 2014, first published 1996); see also Toni Weller, *Information History—An Introduction: Exploring an Emergent Field* (Oxford: Chandos Publishing, 2008): 55–83.

27. Mario Perez-Montoro, *The Phenomenon of Information: A Conceptual Approach to Information Flow* (Lanham, MD: Scarecrow Press, 2007): 3.

28. Weller, *Information History,* 76.

29. Webster, *Theories of the Information Society,* 38–67.

30. Ibid., 106–136; Manuel Castells, especially his trilogy, *The Rise of the Network Society, the Information Age: Economy, Society and Culture* (Oxford: Blackwell, 1996); *The Power of Identity, the Information Age: Economy, Society and Culture* (Oxford: Blackwell, 1997), and *The End of the Millennium, the Information Age: Economy, Society and Culture* (Oxford: Blackwell, 1997).

31. Webster, *Theories of the Information Society,* 149–195.

32. Describes itself as a "multidisciplinary journal that advances our understanding of the relationships between information technology and organizational change," https://tisj.sitehost.iu.edu/ (Accessed May 15, 2021).

33. A. Bose, "Information Resources Management: A Glossary of Terms," *Encyclopedia of Library and Information Science* 41 (New York: Marcel Dekker, 1986): 92.

34. Ronald E. Day, *The Modern Invention of Information: Discourse, History, and Power* (Carbondale: Southern Illinois University Press, 2001): 117.

35. Ibid.

36. Ibid.

37. An American animated sitcom of a family living in some future time that aired in three seasons (1962–63, 1985, 1987); 75 episodes were produced.

38. His seminal text explaining his argument is Samuel P. Huntington, *The Clash of Civilizations and the Remaking of World Order* (New York: Simon & Schuster, 1996).

39. Rebecca Fannin, *Tech Titans of China* (Boston, MA: Nicholas Brealey, 2019); Barry J. Naughton, *The Chinese Economy: Adaptation and Growth* (Cambridge, MA: MIT Press, 2018): 363–394.

40. On Herbert Schiller, see Richard Maxwell, *Herbert Schiller* (Lanham, Md.: Rowman & Littlefield, 2003). Otherwise, his body of work is prodigious; Cortada, *All the Facts.*

41. In particular by Jürgen Habermas, *The Structural Transformation of the Public Sphere: An Inquiry a Category of Bourgeois Society* (Cambridge, MA: MIT Press, 1989); Daniel Headrick, *When Information Came of Age: Technologies of Knowledge in the Age of Reason and Revolution, 1700–1850* (New York: Oxford University Press, 2000).

42. Anthony Giddens, *Social Theory and Modern Sociology* (Cambridge: Polity Press, 1987): 27, 174, 192–193.

43. Edward Higgs, *The Information State in England: The Central Collection of Information on Citizens Since 1500* (Basingstoke: Palgrave Macmillan, 2004), but do not overlook his other book, *Life, Death and Statistics: Civil Registration, Censuses and the Work of the General Register Office, 1836–1952* (Hatfield: University of Hertfordshire Press, 2004).

44. Jon Agar, *The Government Machine: A Revolutionary History of the Computer* (Cambridge, MA: MIT Press, 2003): 15–44; Weller, *Information History*, 80.

45. Weller, *Information History*, 80.

46. Headrick, *When Information Came of Age*, 7–8.

47. Bertram M. Gross, *The Managing of Organizations,* two vols. (New York: Free Press, 1964), especially vol. two, and *Social Intelligence for America's Future: Explorations in Societal Problems* (Boston, MA: Allyn & Bacon, 1969).

48. Neil Postman, *Conscientious Objections: Stirring Up Trouble About Language, Technology and Education* (New York: Knopf, 1988): 162.

49. Alvin Toffler, "The Future as a Way of Life," *Horizon Magazine* 7, no. 3 (Summer 1965): 109–115, and *Future Shock* (New York: Random House, 1970).

50. In *The Third Wave* (New York: William Morrow, 1980), also a "best seller," the author thought of the post-industrial society as being the third wave of civilization, one that would include PCs, something like the Internet, cable television, and cell phones. He spoke about "the intelligent environment" (AI) and "the electronic cottage."

51. Those who studied information overload normally point to computers as the source of this problem for producing so much, although it has been an issue since the eighteenth century when Denis Diderot complained about so many books being published, *Encyclopédie* (1755). For a useful exploration of the issue, see Angela Edmunds and Anne Morris, "The Problem of Information Overload in Business Organizations: A Review of the Literature," *International Journal of Information Management* 20, no. 1 (February 2000): 17–28; also Starr Roxanne Hiltz and Murray Turoff, "Structuring Computer-Mediated Communication Systems to Avoid Information Overload," *Communications of the ACM* 28, no. 7 (July 1985): 680–689; Charles A. O'Reilly, "Individuals and Information Overload in Organizations: Is More Necessarily Better?," *Academy of Management Journal* 23, no. 4 (1980): 684–696

52. Keith Schneider, "Alvin Toffler, Author of 'Future Shock,' Dies at 87," *The New York Times*, June 29, 2010, https://www.nytimes.com/2016/06/30/books/alvin-toffler-author-of-future-shock-dies-at-87.html (Accessed May 15, 2021).

53. Ibid.

54. Ibid.

55. I discuss the French encounters with computing in James W. Cortada, *The Digital Flood: The Diffusion of Information Technology Across the U.S., Europe, and Asia* (New York: Oxford University Press, 2012): 112–124.

56. Jean Jacques Servan-Schreiber, *American Challenge* (New York: Atheneum, 1968), the original edition was *Le Défi Américain* (Paris: Denoël, 1967), and was, too, a national best seller in France and other parts of Europe.

57. Marwick, *The Sixties*, 98, 115, 257, 716, 760–764; Barry Eichengreen, *The European Economy Since 1945* (Princeton, NJ: Princeton University Press, 2007): 103–105, 238–242, 258; Paul Johnson, *Modern Times: From the Twenties to the Nineties* (New York: HarperCollins, 1983): 590–596.

58. David Bell, *Presidential Power in Fifth Republic France* (Oxford: Berg Publishers, 2000); J.R. Frears, *France in the Giscard Presidency* (London: George Allen & Unwin, 1981).

59. He wrote his memoirs, Alain Minc, *Voyage ou Centre du "Système"* (Paris: Bernard Grosset, 2019).

60. In fact, a trade edition appeared immediately, published under the same title by Seuil in a paperback mass-market edition. The official edition under the same title was published in Paris by La Documentation Française, 1978.

61. *The Computerization of Society: A Report to the President of France* (Cambridge, MA: MIT Press, 1980).

62. Anne Mayère, *Pour Une Economie de L'Information* (Paris: Editions de Center National de la Recherche Scientifique, 1990) and Jean Lojkine, *La Révolution Informationnelle* (Paris: Presses Universitaires de France, 1992), are both well stocked with bibliographic references to other debates.

63. Bernard Marti, *Telematics, Techniques, Standards, Services* (Paris: Dunod, 1990); Julien Mailland and Kevin Driscoll, *Minitel: Welcome to the Internet* (Cambridge, MA: MIT Press, 2017): 1–23; Jean-Yves Rincé, *Le minitel* (Paris: Que Sais-Je, Presses Universitaires de France, 1985).

64. Which I discuss in more detail, James W. Cortada, *IBM: The Rise and Fall and Reinvention of a Global Icon* (Cambridge, MA: MIT Press, 2019): 290–291 and in *The Digital Flood*, 112–124.

CHAPTER 4

1. Kenneth Cukier, "Data, Data Everywhere," *The Economist*, Special Report, February 2010, unpaginated, https://www.economist.com/special-report/2010/02/27/data-data-everywhere (Accessed July 28, 2023).

2. *Forbes*, April 2, 2012.

3. For example, in the United States, Mark Jurkowitz, "Americans are Following News about Presidential Candidates Much Less Closely than COVID-19 News," *Pew Research Center*, May 22, 2020, https://www.pewresearch.org/fact-tank/2020/05/22/americans-are-following-news-about-presidential-candidates-much-less-closely-than-covid-19-news/ (Accessed July 28, 2023).

4. The firm admitted its security failure, "Suspending Cambridge Analytics and SCL Group from Facebook," Press Release, Facebook Newsroom, March 16, 2018, https://about.fb.com/news/2018/03/suspending-cambridge-analytica/ (Accessed May 29, 2021); Sarah Frier, "How Facebook Made Its Cambridge Analytica Data Crisis Even Worse," *Bloomberg*, March 19, 2018, https://www.bloomberg.com/news/articles/2018-03-20/how-facebook-made-its-cambridge-analytica-data-crisis-even-worse (Accessed May 29, 2021).

5. Martin Coulter, "Find Out if Your Facebook Data Was Shared with Cambridge Analytica," *Evening Standard*, April 10, 2018, https://www.standard.co.uk/tech/how-to-find-out-if-your-facebook-data-was-shared-with-cambridge-analytica-using-new-tool-launched-by-a3810551.html (Accessed May 29, 2021).

6. "Cambridge Analytica Could Have Also Accessed Private Facebook Messages," *Wired*, April 18, 2018, https://www.wired.com/story/cambridge-analytica-private-facebook-messages/ (Accessed May 29, 2021).

7. Approximately 90 percent of cases included fever, according to the World Health Organization.

8. See, for example its maps of the U.S.A. with color codes showing "hot spots." Fred Bazzoli, "Digital Thermometer Data May Provide Insight into COVID-19 Surges," *HealthcareITNews,* March 26, 2020, https://www.healthcareitnews.com/news/digital-thermometer-data-may-provide-insight-covid-19-surges; Donald G. McNeil, Jr., "Can Smart Thermometers Track the Spread of the Coronavirus?," *New York Times*, March 18, 2020, https://www.nytimes.com/2020/03/18/health/coronavirus-fever-thermometers.html (Both accessed July 28, 2023).

9. Donald G. McNeil, Jr., "Can Smart Thermometers Track the Spread of Coronavirus?," *New York Times*, March 18, 2020.

10. Per Facebook's testimony to the U.S. Congress's Energy and Commerce Committee, June 29, 2018, in a 752 page written document, https://docs.house.gov/meetings/IF/IF00/20180411/108090/HHRG-115-IF00-Wstate-ZuckerbergM-20180411.pdf; for Mark Zuckerberg's personal testimony to the U.S. Senate Committee on Commerce, Science and Transportation, https://www.judiciary.senate.gov/imo/media/doc/04-10-18 percent20Zuckerberg%20Testimony.pdf (Both Accessed July 28, 2023).

11. Chris Wiggins and Matthew L. Jones, *How Data Happened: A History from the Age of Reason to the Age of Algorithms* (New York: W.W. Norton, 2023), quoted on p.5, but see the entire book for well-informed discussions of Big Data.

12. James Bridle, *New Dark Age: Technology and the End of the Future* (New York: Verso, 2018): 84. His book cautions against the potential consequences and limits of information and its technologies, doing so in a thoughtful way.

13. John R. Mashey, "Big Data . . . and the Next Wave of InfraStress," Slides from invited talk, Usenix, April 25, 1998, https://static.usenix.org/event/usenix99/invited_talks/mashey.pdf (Accessed July 28, 2023).

14. Charles Fox, *Data Science for Transport* (Berlin: Springer, 2018): 7.

15. Hamid Ekbia et al., "Big Data, Bigger Dilemmas: A Critical Review," *Journal of the Association for Information Science and Technology* 66, no. 8 (2015): 1525; L. Floridi, "Big Data and Their Epistemological Challenge," *Philosophy and Technology* 25, no. 4 (2012): 435.

16. David Weinberger, *Too Big to Know: Rethinking Knowledge Now That the Facts Aren't the Facts, Experts Are Everywhere, and the Smartest Person in the Room Is the Room* (New York: Basic Books, 2012); Anthony McCosker and Rowan Wilken, "Rethinking "Big Data' as Visual Knowledge: The Sublime and the Diagrammatic in Data Visualization," *Visual Studies* 29, no. 2 (2014): 155–164.

17. Ekbia, "Big Data, Bigger Dilemmas: A Critical Review," 1526.

18. Donna Tam, "Facebook by the Numbers: 1.06 Billion Monthly Active Users," *C/NET*, January 30, 2013, https://www.cnet.com/news/facebook-by-the-numbers-1-06-billion-monthly-active-users/ (Accessed July 28, 2023).

19. Doug Laney, "3D Data Management: Controlling Data Volume, Velocity, and Variety," *Application Delivery Strategies META Group*, February 6, 2001, https://blogs.gartner.com/doug-laney/files/2012/01/ad949-3D-Data-Management-Controlling-Data-Volume-Velocity-and-Variety.pdf (Accessed July 28, 2023).

20. Ekbia et al., "Big Data, Bigger Dilemmas: A Critical Review," 1526.

21. Laney, "3D Data Management: Controlling Data Volume, Velocity, and Variety."

22. Ibid.

23. Ekbia et al., "Big Data, Bigger Dilemmas: A Critical Review," 1526.

24. Ibid., 1527; P.G. Capek, S.P. Frank, S. Gerdt, and D. Shields, "A History of IBM's Open-Source Involvement and Strategy," *IBM Systems Journal* 44, no. 2two (2005): 249–257; Derrick Harris, "The History of Hadoop: From four Nodes to the Future of Data," *GIGAOM*, March 4, 2013, https://gigaom.com/2013/03/04/the-history-of-hadoop-from-4-nodes-to-the-future-of-data/ (Accessed July 28, 2023); R. Kling and S. Iacono, "Computerization Movements and the Mobilization of Support for Computerization," in S.L. Star (ed.), *Ecologies of Knowledge: Work and Politics in Science and Technology* (Albany, NY: State of New York University Press, 1995): 119–153; M. Woldrop, "Wikiomics," *Nature* 455, no. 7290 (January 2008): 22–25.

25. Boyd and Crawford, "Six Provocations for Big Data Community Cleverness Required," *Nature* 455, no. 7209 (September 2008): 1. For a chapter summary of the Vs, Wiggins and Jones, *How Data Happened*, 141–174, which includes a healthy dose of discussion regarding the role of privacy in Big Data.

26. Melanie Feinberg, *Everyday Adventures with Unruly Data* (Cambridge, MA: MIT Press, 2022): 121.

27. For a typical example, IDC, *The Digitization of the World From Edge to Core* (Framingham, MA: IDC, 2018).

28. For example, Ceylan Onay and Elif Ozturk, "A Review of Credit Scoring Research in the Age of Big Data," *Journal of Financial Regulation and Compliance* 26, no. 3 (2018): 382–405; "Big Data's Fourth V," *Spotless*, October 13, 2017, https://spotlessdata.com/blog/big-datas-fourth-v (Accessed July 28, 2023).

29. B.A. Marron and P.A.D. de Maine, "Automatic Data Compression," *Communications of the ACM* 10, no. 11 (November 1967): 710–715, quote p. 710.

30. For a brief, well-written summary, Gil Press, "A Very Short History of Big Data," *Enterprise & Cloud,* May 9, 2013.

31. Hal B. Becker, "Can Users Really Absorb Data at Today's Rates? Tomorrow's?," *Data Communications* 15, no. 8 (1986): 177–193.

32. Peter J. Denning, "The Science of Computing: Saving All the Bits," *American Scientist* 78, no. 5 (September–October 1990): 402–405, quote, p. 403.

33. For example, Michael Cox and David Ellsworth, "Application-controlled Demand Paging for Out-of-core Visualization," *Proceedings of the IEEE 8th Conference on Visualization '97* (October 1997): 235–244.

34. Laney, "3D Data Management: Controlling Data Volume, Velocity, and Variety."

35. Randal E. Bryant, Randy H. Katz, and Edward D. Lazowska, "Big-Data Computing: Creating Revolutionary Breakthroughs in Commerce, Science and Society," white paper, Version 8, December 22, 2008, unpaginated.

36. Ibid.

37. Caleb Scharf, *The Ascent of Information: Books, Bits, Genes, Machines, and Life's Unending Algorithm* (New York: Riverhead Books, 2021).

38. Ibid.

39. Danah Boyd and Kate Crawford, "Critical Questions for Big Data," *Information, Communication & Society* 15, no. 5 (May 2012): 663.

40. Colin Koopman, *How We Be Came Our Data: A Genealogy of the Informational Person* (Chicago, IL: University of Chicago Press, 2019): 178.

41. For example, historian Chad Wellmon, "Knowledge," in Michele Kennerly, Samuel Frederick, and Jonathan E. Abel (eds.), *Information: Keywords* (New York: Columbia University Press, 2021), 133–147.

42. Ekbia et al., "Big Data, Bigger Dilemmas: A Critical Review," 1530.

43. For example, Eric Schatzberg, *Technology: Critical History of a Concept* (Chicago, IL: University of Chicago Press, 2018): 139–140, 230; Neil Postman, *Technopoly: The Surrender of Culture to Technology* (New York: Vintage, 1992): 3–20; P.H. Sawyer and R.H. Hilton, "Technical Determinism: The Stirrup and the Plough," *Past & Present* 24 (April 1963): 90–100; Merritt Roe Smith and Leo Marx (eds.), *Does Technology Drive History? The Dilemma of Technological Determinism* (Cambridge, MA: MIT Press, 1994).

44. D.M. Berry, *The Philosophy of Software: Code and Mediation in the Digital Age* (London: Palgrave Macmillan, 2011): 22.

45. Wiggins and Jones, *How Data Happened*, 225.

46. Borgman, *Scholarship in the Digital Age*, xvii.

47. Daniel E. Atkins et al., *Revolutionizing Science and Engineering Through Cyberinfrastructure: Report of the National Science Foundation Blue-Ribbon Panel on Cyberinfrastructure* (Washington, DC: National Science Foundation, January 2003): 33–47, https://www.nsf.gov/cise/sci/reports/atkins.pdf (Accessed July 29, 2023).

48. Borgman, *Scholarship in the Digital Age*, 6.

49. Ekbia et al., "Big Data, Bigger Dilemmas: A Critical Review," 1530.

50. For example, in 2011–2012, I had access to data extracted from more than 21,000 two-to-three hour interviews conducted with executives from many industries around the world to gain insights on what was on their minds. It took great self-control by me—trained as a historian—not to act like a child in a candy store where the statistician, in effect was saying to me, "you can have as much candy as you want, I will buy it for you and you can have it right now."

51. Summarized in Ekbia et al., "Big Data, Bigger Dilemmas: A Critical Review," 1530–1532.

52. O'Reilly Media, *Big Data Now* (Sebastopol, CA: O'Reilly Media, 2011): 6; David Bollier, *The Promise and Peril of Big Data* (Washington, DC: Aspen Institute, 2010): 13–14.

53. Ekbia et al., "Big Data, Bigger Dilemmas: A Critical Review," 1531. William Bruce Cameron published *Informal Sociology: A Casual Introduction to Sociological Thinking* (New York: Random House, 1963) in which he made the comment, "not everything that can be counted counts, and not everything that counts can be counted," 13.

54. Andrew Gelman, "Too Good To Be True: The Scientific Mass Production of Spurious Statistical Significance," *Slate*, July 24, 2013, https://slate.com/technology/2013/07/statistics-and-psychology-multiple-comparisons-give-spurious-results.html (Accessed July 29, 2023). Gelman concludes his blog with the following: "In one of his stories, science fiction writer and curmudgeon Thomas Disch wrote, 'Creativeness is the ability to see relationships where none exist.' We want our scientists to be creative, but we have to watch out for a system that allows any hunch to be ratcheted up to a level of statistical significance that is then taken as scientific proof."

55. Ekbia et al., "Big Data, Bigger Dilemmas: A Critical Review," 1532.

56. CDC, "COVID-19 Forecasts: Cumulative Deaths," Centers for Disease Control and Prevention, May 28, 2020, https://www.cdc.gov/coronavirus/2019-ncov/covid-data/forecasting-us.html (Accessed July 29, 2023); Nicholas P. Jewell, Joseph A. Lewnard, and Britta L. Jewell, "Predictive Mathematical Models of the COVID-19 Pandemic: Underlying Principles and Value of Projections," *JAMA Network*, April 14, 2020, https://jamanetwork.com/journals/jama/fullarticle/2764824 (Accessed July 29, 2023).

57. Nikolas Rose and Joelle M. Abi-Rached, *Neuro: The New Brain Sciences and the Management of the Mind* (Princeton, NJ: Princeton University Press, 2013): 14.

58. Sarah Lamdan, *Data Cartels: The Companies That Control and Monopolize Our Information* (Stanford, CA: Stanford University Press, 2023).

59. Wiggins and Jones, *How Data Happened*, 150–169.

60. Ibid., 63.

61. Last time I looked, mine was a thirty-three, Nobel Prize winners on average are at about thirty, and forty is off the charts.

62. Daniel Kahneman, Oliver Sibony, and Cass R. Sunstein, *Noise: A Flaw in Human Judgment* (New York: Little, Brown Spark, 2021): 135.

63. "Data, Data Everywhere," *Economist*.

64. Aaron David, "The Government and Big Data: Use, Problems and Potential," *Computerworld*, March 21, 2012, https://www.computerworld.com/article/2472667/the-government-and-big-data--use--problems-and-potential.html (Accessed July 29, 2023).

65. On the challenges posed by Big Data in randomized clinical trials, Vojtech Huser and James J. Cimino, "Impending Challenges for the Use of Big Data," *International Journal of Radiation Oncology, Biology, Physics* 95, no. 3 (July 2016): 890–894; Wullianallur Raghupathi and Viju Raghupathi, "Big Data Analytics in Healthcare: Promise and Potential," *Health Information Science and Systems* 2, no. 1 (December 2014): 3, https://www.ncbi.nlm.nih.gov/pmc/articles/PMC4341817/ (Accessed July 29, 2023).

66. Marshall Allen, "Health Insurers Are Vacuuming Up Details About You—And It Could Raise Your Rates," *ProPublica*, July 17, 2018, https://www.propublica.org/article/health-insurers-are-vacuuming-up-details-about-you-and-it-could-raise-your-rates (Accessed July 29, 2023). This article explains such companies "are tracking your race, education level, TV habits, marital status, net worth. They're collecting what you post on social media, whether you're behind on your bills, what you order online. Then they feed this information into complicated computer algorithms that spit out predictions about how much your health care could cost them."

67. Kevin Ashton, "That 'Internet of Things' Thing," *RFID Journal*, June 22, 2009, https://www.rfidjournal.com/that-internet-of-things-thing (Accessed July 29, 2023); see also "Internet of Things," *Wikipedia*, https://en.wikipedia.org/wiki/Internet_of_things (Accessed May 30, 2020).

68. A surprisingly detailed and up-to-date source is "Big Data," *Wikipedia*, https://en.wikipedia.org/wiki/Big_data (Accessed July 29, 2023). It was updated on May 30, 2021.

69. Borgman, *Scholarship in the Digital Age*, 6.

70. Robert R. Korfhage, *Information Storage and Retrieval* (New York: John Wiley & Sons, 1997): 8–10, passim; Richard K. Belew, *Finding Out About: A Cognitive Perspective on Search Engine Technology and the W.W.W.* (Cambridge: Cambridge University Press, 2000); Michael Lesk, *Understanding Digital Libraries*, 2nd ed. (San Francisco, CA: Morgan Kaufmann, 2005): 375–386.

71. For a link between old and new library and search functions, see Belew, *Finding Out About*.

72. Borgman, *Scholarship in the Digital Age*, 8.

73. Ibid., 9.

74. Ibid.,10.

75. Pablo J. Boczkowski, *Abundance: On the Experience of Living in a World of Information Plenty* (New York: Oxford University Press, 2021): first quote, p. 175, second quote, p. 176.

76. For example, see the complaint even from distinguished scholars, in this instance economist Thomas Piketty, *Capital and Ideology* (Cambridge, MA: Belknap Press of Harvard University Press, 2020): 1040–1041.

77. For a useful discussion of the issue, see Harvey J. Graff, *Undisciplining Knowledge: Interdisciplinarity in the Twentieth Century* (Baltimore, MD: Johns Hopkins University Press, 2015), 129–159.

78. Boyd and Crawford, "Six Provocations for Big Data: Community Cleverness Required," 665.

79. Ibid., 667.

80. Ibid., 667–668; for corroboration, Bollier, *The Promise and Peril of Big Data.*

81. C. Thi Nguyen, "The Limits of Data," *Issues: National Academies of Sciences, Engineering, and Medicine* XL, no. two (Winter 2024), https://issues.org/limits-of-data-nguyen/ (Accessed March 29, 2024).

82. Ibid.

83. Bollier, *The Promise and Peril of Big Data.*

84. Jonathan Vanian, "Why Data is the New Oil," *Fortune*, July 11, 2016; Perry Rotella, "Is Data the New Oil?," *Forbes*, April 11, 2012; Dan Breznitz, *Innovation in Real Places: Strategies for Prosperity in an Unforgiving World* (New York: Oxford University Press, 2021): 175–184.

85. Breznitz, *Innovation in Real Places*, 180.

86. Natasha Singer and Kate Conger, "Google Is Fined $170 Million for Violating Children's Privacy on YouTube," *New York Times,* September 4, 2019.

87. Mike Isaac and Natasha Singer, "Facebook Agrees to Extensive New Oversight as Part of $5 Billion Settlement," Ibid., July 24, 2019.

88. "16 Ways Facebook, Google, Apple and Amazon Are in Government Cross Hairs," Ibid., September 9, 2019, https://www.nytimes.com/interactive/2019/technology/tech-investigations.html (Accessed July 29, 2023). This article has pointers to online versions of the previous several citations to NYT articles.

89. Ibid., 9.

90. Ibid., 10.

91. Ibid.

92. Ibid., 15–16.

93. Derek Thompson, "Google's CEO: 'The Laws Are Written by Lobbyists'," *The Atlantic*, October 1, 2010, https://www.theatlantic.com/technology/archive/2010/10/googles-ceo-the-laws-are-written-by-lobbyists/63908/ (Accessed July 29, 2023).

94. Shoshana Zuboff and James Maxmin, *The Age of Surveillance Capitalism: The Fight for a Human Future at the New Frontier of Power* (New York: PublicAffairs, 2019).

95. Shoshana Zuboff, "You Are Now Remotely Controlled," *New York Times*, January 24, 2020, published in both print and digital editions.

96. Ibid.

97. Ekbia et al., "Big Data, Bigger Dilemmas: A Critical Review," 1540.

98. Ibid.

99. Particularly in James W. Cortada, *The Digital Hand*, vol. II, *How Computers Changed the Work of American Financial, Telecommunications, Media, and Entertainment Industries* (New York: Oxford University Press, 2006).

100. Quoted in Bollier, *The Promise and Peril of Big Data,* 4; Chris Anderson, "The End of Theory: The Data Deluge Makes the Scientific Method Obsolete,"

Wired, June 23, 2008, at http://www.wired.com/science/discoveries/magazine/16-07/pb_theory (Accessed July 29, 2023).

101. Quoted in Bollier, *The Promise and Peril of Big Data,* 4.

102. Ibid., 4–5.

103. Ibid., 5.

104. John Timmer, "Why the Cloud Cannot Obscure the Scientific Method," *Ars Technica,* June 25, 2008, at http://arstechnica.com/old/content/2008/06/why-the-cloud-cannot-obscure-the-scientific-method.ars (Accessed May 31, 2021).

105. Bollier, *The Promise and Peril of Big Data,* 5.

106. Ibid., 7.

107. "Data exhaust" is data left after users have conducted some computing activity, also referred to as "unconventional data," such as geospatial or time-series data that fills Big Data files. For example, in order to transmit a cell phone call for you, your service provider collects data on where your call originated from or where it is going, data that sits in its computer but is of no interest to you or the service provider, merely of momentary value to complete the phone call. That leftover data is data exhaust, and it is the kind Google, Facebook, and others were able to monetize as predictive analytical inputs relevant, say, to advertisers, but that could also be of use to police departments attempting to determine where someone was at a particular time.

108. Bollier, *The Promise and Peril of Big Data,* 8.

109. Ibid., 40. The Aspen attendees were not alone in their concerns; Jon Crowcroft and Adria Gascon, "Analytics Without Tears or Is There a Way for Data to Be Anonymized and Yet Still Useful?," *Computing Edge* (April 2020): 33–39, especially p. 33.

110. Jürgen Renn, *The Evolution of Knowledge: Rethinking Science for the Anthropocene* (Princeton, NJ: Princeton University Press, 2020): 402.

111. Ibid.

112. Zuboff, *The Age of Surveillance,* 99.

113. "Metadata," *Wikipedia,* https://en.wikipedia.org/wiki/Metadata (Accessed July 29, 2023).

114. Richard Gartner, *Metadata: Shaping Knowledge from Antiquity to the Semantic Web* (Berlin: Springer–Verlag, 2016); Marcia Zeng and Jian Qin, *Metadata* (New York: ALA Neal-Schuman, 2016); Steven J. Miller, *Metadata for Digital Collections* (New York: ALA Neal–Schuman, 2020).

115. Stef W. Kight, "Searching for Answers in a Pandemic," *Axios,* April 27, 2020, https://www.axios.com/coronavirus-google-searches-dc472633-33c8-4ab2-97db-0f17c8e3d28d.html (Accessed July 29, 2023).

116. Fiona Harvey, "Lockdowns Trigger Dramatic Fall in Global Carbon Emissions," *The Guardian* May 19, 2020, https://www.theguardian.com/environment/2020/may/19/lockdowns-trigger-dramatic-fall-global-carbon-emissions (Accessed July 29, 2023).

117. Gabriel J.X. Dance and Lazaro Gamio, "As Coronavirus Restrictions Lift, Millions in U.S. Are Leaving Home Again," *New York Times,* May 12, 2020, https://www.nytimes.com/interactive/2020/05/12/us/coronavirus-reopening-shutdown.html (Accessed July 29, 2023).

118. Kelly Kasulis, "S. Korea's Smartphone Apps Tracking Coronavirus Won't Stop Buzzing," *Aljazeera*, April 8, 2020, https://www.aljazeera.com/news/2020/04/korea-smartphone-apps-tracking-coronavirus-won-stop-buzzing-200408074008185.html (Accessed July 29, 2023).

119. Azi Paybarah, Matthew Bloch, and Scott Reinhard, "Where New Yorkers Moved to Escape Coronavirus," *New York Times*, May 16, 2020, https://www.nytimes.com/interactive/2020/05/16/nyregion/nyc-coronavirus-moving-leaving.html (Accessed May 31, 2021).

120. Example of a typical daily report, "Coronavirus in the U.S.: Latest Map and Case Count," Ibid. March 31, 2020, https://www.nytimes.com/interactive/2020/us/coronavirus-us-cases.html (Accessed May 31, 2021).

121. Kahneman, Sibony, and Sunstein, *Noise*, 3–6.

CHAPTER 5

1. Joseph Rain, *The Unfinished Book about Who We Are* (Amazon Kindle/Lucita Publishing, 2018).

2. "President Trump Made 18,000 False or Misleading Claims in 1,170 Days," *Washington Post*, April 14, 2020.

3. "Veracity of Statements by Donald Trump," *Wikipedia*, https://en.wikipedia.org/wiki/Veracity_of_statements_by_Donald_Trump (Accessed June 1, 2023).

4. "Garbage In," *The Economist*, May 30, 2020, p. 77.

5. Deanna Marcum and Roger C. Schonfeld, *Along Came Google: A History of Library Digitization* (Princeton, NJ: Princeton University Press, 2021), 189.

6. Ibid., 210.

7. Thomas Rid, *Active Measures: The Secret History of Disinformation and Political Warfare* (New York: Farrar, Straus, and Giroux, 2020): All quotes, 425–426.

8. Timothy Snyder, *The Road to Unfreedom: Russia, Europe, America* (New York: Crown Publishing, 2018): 159–216 and also *On Tyranny: Twenty Lessons from the Twentieth Century* (New York: Crown, 2017).

9. Harry G. Frankfurt, *On Truth* (Princeton, NJ: Princeton University Press, 2006): 4. This book is a sequel to his more widely discussed book, *On Bullshit* (Princeton, NJ: Princeton University Press, 2005).

10. Frankfurt, *On Truth*, 16–17.

11. Postmodernism essentially distrusts grand theories and ideologies, questioning the role of power relationships and how truths and worldviews are constructed. It is a philosophy that challenges assumptions evident in many disciplines. For more on the topic, see Charles Arthur Willard, *Liberalism and the Problem of Knowledge: A New Rhetoric for Modern Democracy* (Chicago, IL: University of Chicago Press, 1996); Stephen R.C. Hicks, *Explaining Postmodernism: Skepticism and Socialism from Rousseau to Foucault* (Roscoe, IL: Ockham's Razor, 2011); Christopher Butler, *Postmodernism: A Very Short Introduction* (New York: Oxford University Press, 2003).

12. On taking the story back further in time, Peter Burke, *Ignorance: A Global History* (New Haven, CT: Yale University Press, 2023), 197–227.

13. Cases discussed more fully here: James W. Cortada and William Aspray, *Fake News Nation: The Long History of Lies and Misinterpretations in America* (Lanham, MD: Rowman & Littlefield, 2019).

14. For a list of over 150 of his nicknames for domestic and foreign politicians, media figures, organizations, and others, see "List of nicknames used by Donald Trump," *Wikipedia,* https://en.wikipedia.org/wiki/List_of_nicknames_used_by_Donald_Trump (Accessed June 3, 2023).

15. James Harvey Young is the leading authority on this class of information. *The Toadstool Millionaires: A Social History of Patent Medicines in America before Federal Regulation* (Princeton, NJ: Princeton University Press, 1961), *The Medical Messiahs: A Social History of Quackery in Twentieth-Century America* (Princeton, NJ: Princeton University Press, 1967), and *American Health Quackery* (Princeton, NJ: Princeton University Press, 1992).

16. Victor Herbert, *Nutrition Cultism: Facts and Fictions* (Philadelphia, PA: George F. Stickley, 1980).

17. Naomi Oreskes and Erik M. Conway, *Merchants of Doubt: How a Handful of Scientists Obscured the Truth on Issues from Tobacco Smoke to Global Warming* (New York: Bloomsbury, 2010): 10–35; Cortada and Aspray, *Fake News Nation*, 159–173.

18. For an introduction to the practice, see Naomi Oreskes and Erik M. Conway, *The Big Myth: How American Business Taught Us to Loathe Government and Love the Free Market* (New York: Bloomsbury, 2023), 16–119.

19. Elizabeth Cox, Rachael Ann Barry, and Stanton Glantz, "E-cigarette Policymaking by Local and State Governments: 2009–2014," *Milbank Quarterly* 94, no. 3 (September 2016): 520–596, https://www.ncbi.nlm.nih.gov/pmc/articles/PMC5020143/; Johanna Catherine Maclean, Melissa Oney, Joachim Marti, and Jody Sindelar, "What Factors Predict the Passage of State-level E-cigarette Regulations?," *Health Economics* 27, no. 5 (May 2018): 897–907, https://www.ncbi.nlm.nih.gov/pmc/articles/PMC5882548/ (Both accessed June 9, 2023).

20. For example, Mike Magee, *Code Blue: Inside America's Medical Industrial Complex* (New York: Atlantic Monthly Press, 2019).

21. Elizabeth Dias, "The Apocalypse as an "Unveiling': What Religion Teaches Us about the End Times," *New York Times,* April 6, 2020, https://www.nytimes.com/2020/04/02/us/coronavirus-apocalypse-religion.html (Accessed June 9, 2023); Philip Jenkins, *Mystics and Messiahs: Cults and New Religions in American History* (New York: Oxford University Press, 2000): 24–46; Elizabeth Hayes Alvarez, *The Valiant Woman: The Virgin Mary in Nineteenth-Century American Culture* (Chapel Hill: University of North Carolina Press, 2016).

22. To understand how this works, turn to political scientist Murray Edelman's many books, but begin with *Political Language: Words That Succeed and Politics That Fail* (New York: Academic Press, 1977).

23. Discussed in more detail in James W. Cortada and William Aspray, "The Magic of Debunking: Interrogating Fake Facts in the United States Since the

Eighteenth Century," *Library & Information History* 35, no. 3 (August 2019): 144–145.

24. Claire Wardle, "Journalism and the New Information Ecosystem: Responsibilities and Challenges," in Melissa Zimdars and Kembrew McLeod (eds.), *Fake News: Understanding Media and Misinformation in the Digital Age* (Cambridge, MA: MIT Press, 2020): 71.

25. Ibid., 71–85.

26. Cortada and Aspray, *Fake News Nation,* 14–18.

27. The best-researched study of all these plots is currently Philip Shenon, *A Cruel and Shocking Act: The Secret History of the Kennedy Assassination* (New York: Henry Holt, 2013).

28. For a brief history, John Kelly, "Everything Is *Weaponized* Now. This Is a Good Sign for Peace," *Slate*, August 30, 2016, https://slate.com/human-interest/2016/08/how-weaponize-became-a-political-cultural-and-internet-term-du-jour.html (Accessed June 9, 2021).

29. For comparative analysis: Ari Rabin-Havt and Media Matters for America, *Lies, Incorporated: The World of Post-Truth Politics* (New York: Anchor Books, 2016).

30. Murray Edelman, a political scientist, wrote extensively on elections and the language and messages they used. One of the most useful for understanding the issue is his *Politics as Symbolic Action: Mass Arousal and Quiescence* (Madison: University of Wisconsin Press, 1971), and also his, *The Symbolic Uses of Politics* (Urbana: University of Illinois, 1985). Additionally, see Dan F. Hahn, *Political Communication: Rhetoric, Government, and Citizens* (State College, PA: Strata Publishing, 1998).

31. A focus of William Aspray and James W. Cortada, *From Urban Legends to Political Fact-Checking: Online Scrutiny in America, 1990–2015* (Cham: Springer, 2019).

32. For a bibliographic essay on these examples, Cortada and Aspray, *Fake News Facts*, 279–286.

33. Matthew Karp, *This Vast Southern Empire: Slaveholders at the Helm of American Foreign Policy* (Cambridge, MA: Harvard University Press, 2016): 14–15, 67–68; Jay Sexton, *The Monroe Doctrine: Empire and Nation in Nineteenth Century America* (New York: Hill and Wang, 2011): 5–13; and for the twentieth century, James N. Cortada and James W. Cortada, *U.S. Foreign Policy in the Caribbean, Cuba, and Central America* (New York: Praeger, 1985): 41–45.

34. Frank Smyth, *The NRA: The Unauthorized History* (New York: Flatiron Books, 2020): 125–152; Scott Meizer, *Gun Crusaders: The NRA's Culture War* (New York: New York University Press, 2012): 131–170.

35. Aspray and Cortada, *From Urban Legends to Political Fact-Checking.*

36. Burke, *Ignorance*, 226.

37. High-profile fact checkers included Snopes, *Washington Post,* and *New York Times.*

38. Max Read, "How Much of the Internet Is Fake? Turns Out, a Lot of It, Actually," *Intelligencer*, December 26, 2018, https://nymag.com/intelligencer/2018/12

/how-much-of-the-internet-is-fake.html; Amy Watson, "Fake News—Statistics & Facts," *Statista*, May 5, 2020, https://www.statista.com/topics/3251/fake-news/ (both accessed June 8, 2022).

39. This refers to a report produced in the U.S. Department of Defense describing American involvement in Vietnam from 1945 to 1967. The report was leaked to the *New York Times*, which published it. It documented that President L.B. Johnson's administration had lied to the public and to the U.S. Congress about the events and progress of the war. Publication created a large outcry and increased calls for the United States to exit the war. The leaker was Daniel Ellsberg, and his memoirs make for compelling reading about lies and fake facts, *Secrets: A Memoir of Vietnam and the Pentagon Papers* (New York: Viking, 2002); many editions of the papers have been published, a convenient source is George C. Herring (ed.), *The Pentagon Papers: Abridged Edition* (New York: McGraw-Hill, 1993).

40. Stuart Ewen, *PR! A Social History of Spin* (New York: Basic Books, 1996); Roland Marchand, *Creating the Corporate Soul: The Rise of Public Relations and Corporate Imagery in American Big Business* (Berkeley: University of California Press, 1998); Fred Koenig, *Rumors in the Marketplace: The Social Psychology of Commercial Hearsay* (New York: Praeger, 1985); Daniel Pope, *The Making of Modern Advertising* (New York: Basic Books, 1983).

41. Amy Mitchell, Jesse Holcomb, and Rachel Weisel, *State of the News Media 2016* (Pew Research Center, 2016), https://assets.pewresearch.org/wp-content/uploads/sites/13/2016/06/30143308/state-of-the-news-media-report-2016-final.pdf (Accessed June 8, 2022). For examples of fake news and American politics in the period 2016–2018, Mike Wendling, "The (almost) Complete History of 'Fake News'," *BBC Trending*, January 22, 2018, https://www.bbc.com/news/blogs-trending-42724320 (Accessed June 8, 2022).

42. I conducted two tests to see what would happen, one serious and the other humorous. One of my little grandsons was reading a series of adventure books called *Captain Underpants*. I Googled the term and then ordered two volumes from Amazon for him. For days, I then received advertisements for both underwear and books in this series. Some of the advertisements were embedded in articles I read in the *New York Times*.

43. These are tracked by many news organizations along with his original content that he tweets; see "Trump Twitter Archives," http://www.trumptwitterarchive.com/archive (Accessed June 8, 2022).

44. Mike McIntire, Karen Yourish, and Larry Buchanan, "In Trump's Twitter Feed: Conspiracy-Mongers, Racists and Spies," *New York Times*, November 2, 2019, https://www.nytimes.com/interactive/2019/11/02/us/politics/trump-twitter-disinformation.html (Accessed June 8, 2022).

45. "A Brief History of Fake News," *CITS*, https://www.cits.ucsb.edu/fake-news/brief-history; "How is Fake News Spread? Bots, People like You and Microtargeting," Ibid., https://www.cits.ucsb.edu/fake-news/spread (Both accessed June 8, 2022); E. Ferrara, O. Varol, C. Davis, F. Menczer, and A. Flammini, "The Rise of Social Bots," *Communications of the ACM* 59, no. 7 (June 1016): 96–104; C. Watts, "Extremist Content and Russian Disinformation Online: Working with Tech to Find

Solutions," Statement prepared for the U.S. Senate Committee on the Judiciary, Subcommittee on Crime and Terrorism, October 31, 2017, https://www.judiciary.senate.gov/imo/media/doc/10-31-17 percent20Watts%20Testimony.pdf (Accessed June 8, 2022).

46. Aspray and Cortada, *From Urban Legends to Political Fact-Checking*, 41–42.

47. Trisha L. Smith et al., "The Grateful Terrorist: Folklore as Psychological Coping Mechanism," *Voices: The Journal of New York Folklore* 36 (Spring–Summer, 2010): 23–27; see also Cortada and Aspray, *Fake News Nation*.

48. Trisha, "The Grateful Terrorist: Folklore as Psychological Coping Mechanism"; Carl Lindhal, "Faces in the Fire: Images of Terror in Oral Marches and in the Wake of 9/11," *Western Folklore* 68, no. 2/3 (Spring/Summer 2009): 209–234. I personally witnessed such behavior among otherwise rational IBM employees in the 1990s when massive layoffs were occurring all over the company.

49. Aspray and Cortada, *From Urban Legends to Political Fact-Checking*, 44–50.

50. Ibid., 43 for sources.

51. Snopes.com is a fact-checking website, originally known as the Urban Legends Reference Pages, established in 1994. For a useful description with numerous references, see "snopes," *Wikipedia*, https://en.wikipedia.org/wiki/Snopes#References (Accessed June 10, 2022).

52. Aspray and Cortada deposited details on these at https://scholar.colorado.edu/concern/book_chapters/8049g572m (Accessed June 10, 2022).

53. James W. Cortada and Edward Wakin, *Betting on America: Why the U.S. Can Be Stronger After September 11: A Tribute to the American Spirit* (Upper Saddle River, NJ: Financial Times/Prentice Hall, 2002).

54. Aspray and Cortada, *From Urban Legends to Political Fact-Checking*, 43–50.

55. Quoting Michael Wood and Karen M. Douglas, Ibid., 50.

56. Julian E. Barnes, Adam Goldman, and Charlie Savage, "Blaming the Deep State: Officials Accused of Wrongdoing Adopt Trump's Response," *New York Times*, December 18, 2018, https://www.nytimes.com/2018/12/18/us/politics/deep-state-trump-classified-information.html (Accessed June 10, 2022); James B. Stewart, *Deep State: Trump, the FBI, and the Rule of Law* (New York: Penguin, 2019).

57. Kurtis Hagen, "Conspiracy Theories and Stylized Facts," *Journal for Peace and Justice Studies* twenty-one (2011): 3–22; and a classic text on the issue, Richard Hofstadter, "The Paranoid Style in American Politics," *Harper's Magazine* (1964): 77–86, also his *The Paranoid Style in American Politics and Other Essays* (Cambridge, MA: Harvard University Press, 1952, 1954, 1964, 1965).

58. Aspray and Cortada, *From Urban Legends to Political Fact-Checking*, 51–55.

59. On how Snopes does its work, Ibid., 55–64.

60. David Mikkelson, "Billy Graham's Daughter's Speech," originally published October 3, 2001, updated March 9, 2018, http://www.snopes.com/rumors/wheregod.asp (Accessed June 10, 2022).

61. Aspray and Cortada, *From Urban Legends to Political Fact-Checking*, 59.

62. Barbara Mikkelson, "Osama bin Laden Captured," last updated May 1, 2011, http://www.snopes.com/rumors/captured.asp (Accessed June 10, 2021).

63. Aspray and Cortada, *From Urban Legends to Political Fact-Checking*, 107.

64. Stefan Wojcik, Adam Hughes, and Emma Remy, "About One-In-Five Adult Twitter Users in the U.S. Follow Trump," *Pew Research Center,* July 15, 2019, https://www.pewresearch.org/fact-tank/2019/07/15/about-one-in-five-adult-twitter-users-in-the-u-s-follow-trump/ (Accessed June 10, 2022).

65. Aspray and Cortada, *From Urban Legends to Political Fact-Checking*, 108.

66. Politifact.com, and for a table listing these cases, Aspray and Cortada, *From Urban Legends to Political Fact-Checking*, 116.

67. For a detailed analysis of the pants controversy, Bill Chappell, "Actually, Trump Was NOT Wearing Hi Pants Backward at a Weekend Rally," *NPR*, June 7, 2021, https://www.npr.org/2021/06/07/1003916275/trump-pants-backward-fact-check (Accessed June 19, 2022).

68. "The Nobel Peace Prize 2021," https://www.nobelprize.org/prizes/peace/2021/summary/ (Accessed January 7, 2022).

69. Maria Ressa, *How to Stand Up to a Dictator: The Fight for Our Future* (New York: HarperCollins, 2022), 183.

70. Although it stimulated some controversy, this is the original study: J. Giles, "Internet Encyclopedias Go Head to Head: Jimmy Wales' Wikipedia Comes Close to Britannica in Terms of Accuracy of Its Science Entries," *Nature* 438, no. 7070 (2005): 900–901; for a detailed discussion, see "Reliability of Wikipedia," *Wikipedia,* https://en.wikipedia.org/wiki/Reliability_of_Wikipedia#cite_note-GilesJ2005Internet-21 (Accessed June 10, 2022). It cites multiple studies concerning this website's factual accuracy. In 2010, a study published by librarians approved of its quality, Barry X. Miller, Karl Helicher, and Teresa Berry, "I Want My Wikipedia!," *Library Journal*, Digitized Edition, May 21, 2010, https://www.libraryjournal.com/?detailStory=i-want-my-wikipedia (Accessed June 10, 2022).

71. Shane Greenstein, "Revenge Editing and Wikipedia," *Digitopoly*, May 26, 2013, https://digitopoly.org/2013/05/26/revenge-editing-and-wikipedia/ (Accessed June 10, 2021).

72. Cambridge Dictionary, https://dictionary.cambridge.org/us/dictionary/english/scrutiny (Accessed June 10, 2022); for further discussion, Aspray and Cortada, *From Urban Legends to Political Fact-Checking,* 1–8.

73. Galen Stocking, "Social Media Bots Draw Public's Attention and Concern," *Pew Research Center,* October 15, 2018. With bots spreading rapidly and becoming more effective (i.e., realistic), they represent a growing challenge for truth seekers. Chengcheng Shao, Giovanni Luca Ciampaglia, Onur Varol, Kai-Cheng Yang, Alessandro Flammini, and Filippo Menczer, "The Spread of Low-Credibility Content by Social Bots," *Nature Communications*, November 20, 2018, https://www.nature.com/articles/s41467-018-06930-7?utm_source=newsletter&utm_medium=email&utm_campaign=newsletter_axiosam&stream=top (Accessed June 10, 2022).

74. Janna Anderson and Lee Rainie, *The Future of Truth and Misinformation Online* (Pew Research Center, October 19, 2017), https://www.pewresearch.org/internet/2017/10/19/the-future-of-truth-and-misinformation-online/ (Accessed June 10, 2022).

75. "Knowledge and Perception Surrounding COVID-19," *Pew Research Center,* March 18, 2020, https://www.pewresearch.org/?s=fake+facts&days=180#recent-publications (Accessed June 10, 2022).

76. Lachlan Cartwright, Maxwell Tani, and Lloyd Grove, "New York Times Executives Take Turns Apologizing to Quell Staff Revolt," *Daily Beast,* June 5, 2020, https://www.thedailybeast.com/new-york-times-executives-take-turns-apologizing-to-quell-staff-revolt-over-tom-cottons-send-in-the-troops?ref=home; Erik Wemple, "A Crisis of Conviction at the New York Times," *Washington Post,* June 5, 2021, file:///Users/JimMac/Downloads/A%20crisis%20of%20conviction%20at%20the%20New%20York%20Times%20-%20The%20Washington%20Post%20(1).pdf (Both accessed June 10, 2022).

77. Pew Research Center, https://www.pewresearch.org/?s=fake+facts&days=180#recent-publications (Accessed June 20, 2022).

78. Anderson and Rainie, "The Future of Truth and Misinformation Online."

79. Ibid.

80. Ibid.

81. Ibid.

82. Laura Gurak, *Cyberliteracy: Navigating the Internet with Awareness* (New Haven, CT: Yale University Press, 2001), is one of the earliest publications on this subject.

83. It still seemed still necessary to advocate for and explain computer literacy; Michael J. Halvorson, *Code Nation: Personal Computing and the Learn to Program Movement in America* (New York: ACM Press, 2020): 99–126; Faithe Wempen, *Digital Literacy for Dummies* (Hoboken, NJ: John Wiley & Sons, 2015): 9–24.

84. Ibid., 12–14.

85. Nicole A. Cooke, *Fake News and Alternative Facts: Information Literacy in a Post-Truth Era* (Chicago, IL: ALA Editions, 2018): especially p. 20.

86. Cortada and Aspray, *Fake News Nation.*

87. Cooke, *Fake News and Alternative Facts,* 10–11.

88. Amy Mitchell, Jeffrey Gottfried, Michael Barthel, and Naomi Sumida, "Distinguishing Between Factual and Opinion Statements in the News," June 18, 2018, Pew Research Center, unpaginated.

89. Ibid.

90. Aspray and Cortada, *From Urban Legends to Political Fact-Checking.*

91. Mark Shepard, *There Are No Facts: Attentive Algorithms, Extractive Data Practices, and the Quantification of Everyday Life* (Cambridge, MA: MIT Press, 2022).

92. I have explored this theme in considerable detail in James W. Cortada, *IBM: The Rise and Fall and Reinvention of a Global Icon* (Cambridge, MA: MIT Press, 2019).

93. For example, Daniel J. Levitin, *A Field Guide to Lies: Critical Thinking in the Information Age* (New York: Dutton, 2016).

94. Cooke, *Fake News and Alternative Facts,* 18.

95. On the concept, see Thomas P. Mackey, *Metaliteracy: Reinventing Information Literacy to Empower Learners* (Chicago, IL: American Library Association, 2014).

CHAPTER 6

1. Jeff Gauthier, Antony T. Vincent, Steve J. Charette, and Nicolas Drome, "A Brief History of Bioinformatics," *Briefings in Bioinformatics* 20, no. 6 (November 2019): 1992.

2. For a useful history of AI, Chris Wiggins and Matthew L. Jones, *How Data Happened: A History from the Age of Reason to the Age of Algorithms* (New York: W.W. Norton, 2023), pp. 124–140.

3. Harry G. Frankfurt, *On Bullshit* (Princeton, NJ: Princeton University Press, 2005).

4. However, for a tightly written, informed review of theories and perspectives, although slightly dated, and worth reading, see Karin Knorr Cetina, *Epistemic Cultures: How the Sciences Make Knowledge* (Cambridge, MA: Harvard University Press, 1999): 5–8.

5. Steven Sloman and Philip Fernbach, *The Knowledge Illusion: Why We Never Think Alone* (New York: Riverhead/Penguin, 2017): 4–6, 10–15.

6. Paul B. Armstrong, *Stories and the Brain: The Neuroscience of Narrative* (Baltimore, MD: Johns Hopkins University Press, 2020); Bob Garrett and Gerald Hough, *Brain & Behavior: An Introduction to Behavioral Neuroscience*, 5th ed. (Thousand Oaks, CA: SAGE, 2018); Bret Stetka, *A History of the Human Brain: From the Sea Sponge to CRISPR, How Our Brains Evolved* (Portland, OR: Timber Press, 2021); Steven Sloman and Philip Fernbach, *The Knowledge Illusion: Why We Never Think Alone* (New York: Riverhead Books/Penguin, 2017).

7. Sloman and Fernbach, *The Knowledge Illusion,* 96–101.

8. S. Banbury and S. Tremblay, *A Cognitive Approach to Situation Awareness: Theory and Application* (Aldershot: Ashgate Publishing, 2004); H. Artman, "Team Situation Assessment and Information Distribution," *Ergonomics* 43, no. 8 (2000): 1111–1128.

9. Jonathan Rauch, *The Constitution of Knowledge: A Defense of Truth* (Washington, DC: Brookings Institution, 2021): 20–25.

10. Jonathan Haidt, *The Righteous Mind: Why Good People Are Divided by Politics and Religion* (New York: Pantheon, 2012).

11. Rauch, *The Constitution of Knowledge,* 24–25.

12. Jim Al-Khalili, *The Joy of Science* (Princeton, NJ: Princeton University Press, 2022): 114–129.

13. Ben Yagoda, "The Cognitive Biases Tricking Your Brain," *The Atlantic,* September 2018, https://www.theatlantic.com/magazine/archive/2018/09/cognitive-bias/565775/ (Accessed July 17, 2021).

14. Pamela Fuller, Mark Murphy, and Anne Chow, *The Leader's Guide to Unconscious Bias: How to Reframe Bias, Cultivate Connection, and Create*

High-Performance Teams (New York: Simon & Schuster, 2020): 17–72; Erin Beeghly and Alex Madva (eds.), *An Introduction to Implicit Bias: Knowledge, Justice, and the Social Mind* (London: Routledge, 2020); Howard J. Ross, *Everyday Bias: Identifying and Navigating Unconscious Judgments in Our Daily Lives* (Lanham, MD: Rowman & Littlefield, 2020): 1–16, 41–60, 85–102, see also his bibliography, 173–180.

15. Rauch, *The Constitution of Knowledge,* 27.

16. Lee McIntyre, *Post-Truth* (Cambridge, MA: MIT Press, 2018).

17. Rauch, *The Constitution of Knowledge,* 27.

18. Jennifer Kavanaugh and Michael D. Rich, *Truth Decay: An Initial Exploration of the Diminishing Role of Facts and Analysis in American Public Life* (Santa Monica, CA: RAND, 2018), available at https://www.rand.org/pubs/research_reports/RR2314.html; see also Jason Altmire, *Dead Center: How Political Polarization Divided America and What to Do About It* (Mechanicsburg, PA: Sunbury Press, 2017); Gary Klein, "The Curious Case of Confirmation Bias," *Psychology Today*, May 5, 2019, and his "Escaping from Fixation," Ibid., June 11, 2019.

19. Your author has been a historian for a half century and, in that period of time, has read much written by fellow historians about truth, but the majority was based on philosophy and research practices, not on neuroscience. One useful exception is Jonathan Gorman, "Where Neuroscience Gets Things Wrong (And Right)," *History and Theory* 58, no. 3 (September 2019): 483–495.

20. Jay J. Van Bavel and Andrea Pereira, "The Partisan Brain: An Identity-Based Model of Political Belief," *Trends in Cognitive Sciences* 22, no. 3 (March 2018): 213–224; Dan M. Kahan, Hank Jenkins-Smith, and Donald Braman, "Cultural Cognition of Scientific Consensus," *Journal of Risk Research* (2010): 1–28; Dan M. Kahan, "Misconceptions, Misinformation, and the Logic of Identity-Protective Cognition," *Cultural Cognition Project Working Paper Series No. 164,* Yale Law School, June 2017, https://papers.ssrn.com/sol3/papers.cfm?abstract-id=2973067 (Accessed July 17, 2021).

21. Rauch, *The Constitution of Knowledge*, 33–37.

22. Pippa Norris, "Closed Minds? Is a "Cancel Culture' Stifling Academic Freedom and Intellectual Debate in Political Science?," *SSRN Electronic Journal* (August 15, 2020), https://papers.ssrn.com/sol3/papers.cfm?abstract_id=3671026 (Accessed August 8, 2021).

23. James W. Cortada and William Aspray, *Fake News Nation: The Long History of Lies and Misinterpretations in America* (Lanham, MD: Rowman & Littlefield, 2019): 73–88.

24. Sloman and Fernbach, *The Knowledge Illusion*, 216.

25. Pablo J. Boczkowski and Eugenia Mitchelstein, *The Digital Environment: How We Live, Learn, Work, and Play Now* (Cambridge, MA: MIT Press, 2021): ix.

26. James W. Cortada, *Building Blocks of Society: History, Information Ecosystems, and Infrastructures* (Lanham, MD: Rowman & Littlefield, 2021).

27. Colin Koopman, *How We Be came Our Data: A Genealogy of the Informational Person* (Chicago, IL: University of Chicago Press, 2019): 177.

28. James W. Cortada, *All the Facts: A History of Information in the United States since 1870* (New York: Oxford University Press, 2016) and *Birth of Modern Facts:*

How the Information Revolution Transformed Academic Research, Governments, and Businesses (Lanham, MD: Rowman & Littlefield, 2023).

29. For introductions to the topic, see Luciano Floridi, *Information: A Very Short Introduction* (New York: Oxford University Press, 2010); Toni Weller, *Information History in the Modern World: Histories of the Information Age* (New York: Palgrave, 2011) and her earlier study, *Information History—An Introduction: Exploring an Emerging Field* (New York: Neal-Schuman, 2008).

30. Paul N. Edwards, *A Vast Machine: Computer Models, Climate Data, and the Politics of Global Warming* (Cambridge, MA: MIT Press, 2013); Kristine C. Harper, *Weather by the Numbers: The Genesis of Modern Meteorology* (Cambridge, MA: MIT Press, 2008).

31. Nick Braisby and Angus Gellatly, *Cognitive Psychology* (New York: Oxford University Press, 2012); Dan H. Sanes, Thomas A. Reh, and William A. Harris, *Development of the Nervous System,* 2nd ed. (Waltham, MA: Academic Press, 2005).

32. For example, Anand Mohan, "The Beauty of Fractal Patterns in Geology," *Journal of Scientific Research* 65, no. 3 (2021): 7–9; Ursula K. Heise, "Unnatural Ecologies: The Metaphor of the Environment in Media Theory," *Configurations* 10, no. 1 (Winter 2002): 149–168.

33. For an example of reporting on this subject, see Peter Wohlleben, *The Hidden Life of Trees: What They Feel, How They Communicate* (Vancouver: Greystone Books, 2015).

34. Both quotes, Bonnie A. Nardi and Vicki L. O'Day, *Information Ecologies: Using Technology with Heart* (Cambridge, MA: MIT Press, 1999): 49.

35. For a discussion of the history of genetics, see Theodore M. Porter, *Genetics in the Madhouse: The Unknown History of Human Heredity* (Princeton, NJ: Princeton University Press, 2018).

36. I did, Cortada, *Building Blocks of Society.*

37. Tufte reminds us, too, that "To choose a model is to choose assumptions—unknown, unseen, forgotten. Some assumptions are worse than others," Edward Tufte, *Seeing with FreshEyes,Meaning, Space, Data, Truth* (Cheshire, CT: Graphics Press, 2020): 8.

38. Jonathan Haidt, *The Righteous Mind: Why Good People Are Divided by Politics and Religion* (New York: Pantheon, 2012): 90.

39. Koopman, *How We Became Our Data*, ix.

40. Rauch, *The Constitution of Knowledge*, 85–87.

41. Helen E. Longino, *Science as Social Knowledge: Values and Objectivity in Scientific Inquiry* (Princeton, NJ: Princeton University Press, 1990), and on her views about how information is socialized, *The Fate of Knowledge* (Princeton, NJ: Princeton University Press, 2002), 97–123.

42. Nardi and O'Day, *Information Ecologies*, 65.

43. Computer vendors, such as IBM, continuously published dictionaries of such phrases in various languages. See, for example, one of the most widely consulted ones, IBM, *IBM Dictionary of Computing* (New York: IBM Corporation, 1994), which appeared in an edition of 758 pages. It was not the first edition, either.

44. Explained in detail in Cortada, *OUP2.*

45. Jacques Ellul, *The Technological Society* (New York: Vintage, 1964): 90–91.

46. Ibid., 91.

47. Published in John von Neumann, "First Draft of a Report on the EDVAC," *IEEE Annals of the History of Computing* 15, no. 4 (1993): 27–76; M.D. Godfrey and D.F. Hendry, "The Computer as von Neumann Planned It," Ibid., 15, no. 1 (1993): 11–21.

48. Chris Wiggins and Matthew L. Jones, *How Data Happened: A History from the Age of Reason to the Age of Algorithms* (New York: W.W. Norton, 2023), 76; but see ibid for a general history of IQ testing, 67–77.

49. Ibid., 76.

50. For a detailed account, see Koopman, *How We Became Our Data*, 66–107.

51. Sloman and Fernbach, *The Knowledge Illusion*, 203.

52. Luria, *Cognitive Development*, 163.

53. James R. Flynn, *What Is Intelligence? Beyond the Flynn Effect* (Cambridge: Cambridge University Press, 2007).

54. Ibid., 114.

55. James R. Flynn, *Asian Americans: Achievement Beyond IQ* (Hillsdale, NJ: Erlbaum, 1991).

56. Ibid., 120. Note also, in addition to his formal research on Chinese-Americans, Flynn was raised in an Irish-American family, so his quip was probably partially informed by his tacit knowledge of Irish culture.

57. Ibid., 122.

58. Ibid., 144.

59. Ibid., 145.

60. Ibid., 175.

61. Howard Gardner, *Multiple Intelligences* (New York: Basic Books, 1993, 2006) and *Frames of Mind: The Theory of Multiple Intelligences* (New York: Basic Books, 1983, 2004, 2011).

62. Historical interlude: Every time, it seems, that a new concept is developed in computing, the computer scientists and engineers give it a label that makes little sense to the masses. Regenerative AI is an example, as were such earlier ones as "relational databases," "concatenation," and "cognitive computing," until someone came along and provided a more understandable label, such as "Big Data."

63. J.D. Mayer, P. Salovey, and D.R. Caruso, "Emotional Intelligence: Theory, Findings, and Implications," *Psychological Inquiry* 15, no. 3 (July 2004): 197–215.

64. On early discussions, see M. Beldoch, "Sensitivity to Expression of Emotional Meaning in Three Modes of Communication," in J.R. Davitz et al. (eds.), *The Communication of Emotional Meaning* (New York: McGraw-Hill, 1964): 31–42; B. Leuner, "Emotional Intelligence and Emancipation," *Praxis der Kinderpsychologie und Kinderpsychiatrie* 15 (1966): 193–203. The literature on EI is now substantial.

65. Daniel Goleman, *Emotional Intelligence: Why It Can Matter More Than IQ* (New York: Random House, 2005).

66. What is a "Renaissance Man" that one hears of, such as Benjamin Franklin, master of diverse bodies of information? They were masters of far smaller collections of diverse information than are available today.

67. Tetlow, *The Web's Awake*, 155–156.

68. George Dyson, *Analogia: The Emergence of Technology Beyond Programmable Control* (New York: Farrar, Straus and Giroux, 2020): 7.

69. Daniel Kahneman, *Thinking, Fast and Slow* (New York: Farrar, Strauss and Giroux, 2011).

70. Richard H. Thaler and Cass R. Sunstein, *Nudge: The Final Edition* (New York: Penguin, 2009, 2nd ed. 2021), which, like Kahneman's study, discusses how to use biases to alter behavior and thinking.

71. David Epstein, *Range: Why Generalists Triumph in a Specialized World* (New York: Riverhead Books/Penguin, 2019): 55–214. I commented on this phenomenon too in James W. Cortada, *History Hunting: A Guide for Fellow Adventurers* (Armonk, NY: M.E. Sharpe, 2012): 223–247, but see also Richard E. Neustadt and Ernest R. May, *Thinking in Time: The Uses of History for Decision Makers* (New York: Free Press, 1986); Gordon C. Wood, *The Purpose of the Past: Reflections on the Uses of History* (New York: Penguin, 2009).

72. For example, tracking data from the *Washington Post*, August 11, 2021, https://www.washingtonpost.com/graphics/2020/national/coronavirus-us-cases-deaths/, Raj Chetty, John N. Friedman, and Michael Stepner, "Who Spent Their Last Stimulus Checks?" *New York Times*, February 8, 2021, https://www.nytimes.com/interactive/2021/02/08/opinion/stimulus-checks-economy.html?action=click&module=Opinion&pgtype=Homepage, Christopher Ingraham, "Researchers Identify Social Factors Inoculating Some Communities Against Coronavirus," *Washington Post*, February 11, 2021 (all Accessed August 11, 2021).

73. For the evidence underpinning this paragraph, Michael Park, Erin Leahey, and Russell J. Funk, "Papers and Patents Are Becoming Less Disruptive Over Time," *Nature* 613 (2023): 138–144 (Accessed January 6, 2023).

74. Jim Al-Khalili, *The World According to Physics* (Princeton, NJ: Princeton University Press, 2020): 241.

75. Caleb Scharf, *The Ascent of Information: Books, Bits, Genes, Machines, and Life's Unending Algorithm* (New York: Riverhead Books, 2021): 284.

76. Jason Steinhauer, *History, Disrupted: How Social Media and the World Wide Web Have Changed the Past* (Cham: Palgrave Macmillan, 2022): 110.

CHAPTER 7

1. Harry G. Frankfurt, *On Truth* (Princeton, NJ: Princeton University Press, 2017): 34.

2. The literature is vast, beginning with two studies: James Suzman, *Work: A Deep History, from the Stone Age to the Age of Robots* (New York: Penguin, 2021); Martin Ford, *The Rise of the Robots: Technology and the Threat of a Jobless Future* (New York: Basic Books, 2015).

3. Rauch, *The Constitution of Knowledge*, 126–138. For a detailed study of this issue, see William Aspray and James W. Cortada, *From Urban Legends to Political Fact-Checking: Online Scrutiny in America, 1990–2015* (Cham: Springer, 2019).

4. In my last decade at IBM (2002–2012) I participated in numerous meetings where the status of AI as a technology and a force and opportunity to be reckoned with in the market was discussed at some of the highest levels in the firm.

5. Steve Lohr, "What Ever Happened to IBM's Watson?," *New York Times*, July 16, 2021, https://www.nytimes.com/2021/07/16/technology/what-happened-ibm-watson.html?action=click&module=In%20Other%20News&pgtype=Homepage (Accessed July 16, 2017). Lohr is a highly regarded expert reporter on IT subjects.

6. Ibid.

7. Ibid.

8. Early in the Watson medical hype, I talked with an IBM computer scientist friend who told me they were not close to being able to diagnose cancer cures, even though he and his colleagues had worked in collaboration with the Mayo Clinic as far back as the 1990s on breast cancer diagnostics that proved useful to doctors.

9. Chris Wiggins and Matthew L. Jones, *How Data Happened: A History from the Age of Reason to the Age of Algorithms* (New York: W.W. Norton, 2023): 128.

10. James Bridle, *The New Dark Age: Technology and the End of the Future* (New York: Verso, 2018): 248.

11. It seems every major news media discussed the topic, but for example, "Free Exchange: Amazing Inventions," *The Economist*, February 4, 2023, p. 66; Charles Seife, "A.I. Like ChatGPT Is Revealing the Insidious Disease at the Heart of Our Scientific Process," *Slate*, January 31, 2023, https://slate.com/technology/2023/01/ai-chatgpt-scientific-literature-peer-review.html (Accessed January 31, 2023); Mike Allen, "1 Big Thing: Tech Giants Rush to Exploit Chatbox," *AXIOS AM*, February 6, 2023, https://www.axios.com/2023/02/06/chatgpt-tech-giants-generative-ai (Accessed February 7, 2023).

12. Ryan Heath, "Western Countries Are More Pessimistic About AI," *AXIOS*, March 12, 2024, https://www.axios.com/2024/03/12/western-countries-ai-generative-ai-productivity (Accessed March 30, 2024).

13. Steven Sloman and Philip Fernbach, *The Knowledge Illusion: Why We Never Think Alone* (New York: Riverhead Books/Penguin, 2017), 140.

14. Rob Thomas and Paul Zikopoulos, *The AI Ladder: Accelerate Your Journey to AI* (Sebastopol, CA: O'Reilly, 2020): 27–28.

15. Ibid., 29.

16. Jennifer A. Kingston, ""AI native' Gen Zers are Comfortable on the Cutting Edge," *Axios*, February 10, 2024.

17. Judith Pintar and David Hopping, *Information Science: The Basics* (London: Routledge, 2023): 157.

18. Harvey J. Graff, *Undisciplining Knowledge: Interdisciplinarity in the Twentieth-Century* (Baltimore, MD: Johns Hopkins University Press, 2015), 122; see Ibid for a useful history of OR, 91–123.

19. For tutoring on three major AI tools as of 2024, see https://www.axios.com/2024/03/03/chatgpt-copilot-gemini-ai-beginners-guide-tools?utm_source=newsletter&utm_medium=email&utm_campaign=newsletter_axiosam&stream=top (Accessed March 30, 2024).

20. Alison Snyder, "Axios Science," *AXIOS*, March 14 and 15, 2024, https://www.axios.com/2024/03/15/artificial-intelligence-neuroscience-brain-body (Accessed March 30, 2024).

21. Norbert Wiener, *God and Golem, Inc.* (Cambridge, MA: MIT Press, 1964): 69.

22. George Dyson, *Analogia: The Emergence of Technology Beyond Programmable Control* (New York: Farrar, Straus and Giroux, 2020): 245.

23. Jennifer A. Kingston, "Coming Soon: A Programmable Army of Humanoid Robots," *AXIOS*, March 14, 2024, https://www.axios.com/2024/03/15/artificial-intelligence-neuroscience-brain-body (Accessed March 30, 2024).

For Further Reading

This bibliographic essay provides a useful, but curated, selection of materials for further reading. It favors books that take a largely historical orientation and are readily available in English. These are organized by chapter topics for convenience. For specific citations to articles and online sources, consult the endnotes. More detailed bibliographies are available in James W. Cortada, *Birth of Modern Facts: How the Information Revolution Transformed Academic Research, Governments, and Businesses* (Lanham, Md.: Rowman & Littlefield, 2023) and in Judith Pintar and David Hopping, *Information Science: The Basics* (London: Routledge, 2023).

LEARNING FROM THE HISTORY OF INFORMATION

I have argued that a useful place to begin is by understanding how the brain processes information, especially since we have learned a great deal about that activity in just the past four decades. Useful introductions to the subject include Peter J. Bowler, *Evolution: The History of an Idea* (Berkeley: University of California Press, 2003); Robert Bud, *The Uses of Life: A History of Biotechnology* (Cambridge: Cambridge University Press, 1993); Kevin Davies, *Cracking the Genome: Inside the Race to Unlock Human DNA* (New York: Free Press, 2001); Stephen Jay Gould, *The Structure of Evolutionary Theory* (Cambridge, Mass.: Belknap Press of Harvard University Press, 2002); Lois M. Magner, *A History of the Life Sciences* 3rd ed. (New York: Marcel Dekker, 2002); Jan Sapp, *Genesis: The Evolution of Biology* (New York: Oxford University Press, 2003); and Steve Sloman and Philip Fernbach, *The Knowledge Illusion: Why We Never Think Alone* (New York: Penguin, 2017). For a discussion linking biological considerations to the operation of organizations and human work, there is the well-informed, easy-to-read classic by Margaret J. Wheatley, *Leadership and the New Science: Discovering Order in a Chaotic World* (San Francisco, Cal.: Berrett-Koehler, 2006). For an even broader study written by a scientist, see Caleb Scharf, *The Ascent of Information: Books, Bits, Genes, Machines, and Life's Unending Algorithm* (New York: Riverhead Books, 2021).

For a more traditional history of information, two thoughtful studies are Luciano Floridi, *Information: A Very Short Introduction* (New York: Oxford University Press, 2010) and Toni Weller, *Information History—an Introduction: Exploring an Emerging Field* (New York: Neal-Schuman, 2008). I have also commented on the topic from an historical perspective in James W. Cortada, *All the Facts: A History of Information in the United States Since 1870* (New York: Oxford University Press, 2016). For a useful, excellent companion study written by information experts, see Judith Pintar and David Hopping, *Information Science: The Basics* (London: Routledge, 2023). My book takes the story down to the late twentieth century, while the Pintar and Hopping book focuses largely on developments in the twenty-first century, but with historical context too. For an explanation of scientific thinking, you can do no better than to read Jonathan Rauch, *The Constitution of Knowledge: A Defense of Truth* (Washington, D.C.: Brookings Institution Press, 2021). But also look at Helen E. Longino, *Science as Social Knowledge: Values and Objectivity in Scientific Inquiry* (Princeton, N.J.: Princeton University Press, 1990), and for her views about how information is socialized, *The Fate of Knowledge* (Princeton, N.J.: Princeton University Press, 2002). For a broad discussion of how information is changing society, but not necessarily for the good, see James Bridle, *New Dark Age: Technology and the End of the Future* (new York: Verso, 2018). For a broad general, more positive historical account that is quite useful, there is Chris Wiggins and Matthew L. Jones, *How Data Happened: A History from the Age of Reason to the Age of Algorithms* (New York: W.W. Norton, 2023). This latter book is a useful companion volume to Cortada, *Birth of Modern Facts*.

HOW INFORMATION CHANGED WORK

Useful for our immediate study of work and information and to explore the implications of more digital technologies and information can be accessed in publications influential on management when they appeared. The diffusion of computing and digital information in workspaces has been studied by historians. Sources include James W. Cortada, *The Digital Flood: The Diffusion of Information Technology Across the U.S. Europe, and Asia* (New York: Oxford University Press, 2012); Martin Campbell-Kelly, William Aspray,

Nathan Ensmenger, and Jeffrey R. Yost, *Computer: A History of the Information Machine* 3rd ed. (Boulder, Col.: Westview, 2014); Thomas Haigh and Paul E. Ceruzzi, *A New History of Modern Computing* 3rd ed. (Cambridge, Mass.: MIT Press, 2021); Martin Campbell-Kelly and Daniel D. Garcia-Swartz, *From Mainframes to Smartphones: A History of the International Computer Industry* (Cambridge, Mass.: Harvard University Press, 2015) and their *Cellular: An Economic and Business History of the International Mobile-Phone Industry* (Cambridge, Mass.: MIT Press, 2022); James W. Cortada, *Before the Computer: IBM, NCR, Burroughs, and Remington Rand and the Industry They Created, 1865–1956* (Princeton, N.J.: Princeton University Press, 1993), *The Computer in the United States: From Laboratory to Market, 1930–1960* (Armonk, N.Y.: M.E. Sharpe, 1993), and *Information Technology as Business History: Issues in the History and Management of Computers* (Westport, CT: Greenwood Press, 1996); JoAnne *Yates, Structuring the Information Age: Life Insurance and Technology in the Twentieth Century* (Baltimore, Md.: Johns Hopkins University Press, 2008). On the System 360, see Emerson W. Pugh, Lyle R. Johnson, and John H. Palmer, *IBM's 360 and Early 370 Systems* (Cambridge, Mass.: MIT Press, 1991) and James W. Cortada, *IBM: The Rise and Fall and Reinvention of a Global Icon* (Cambridge, Mass.: MIT Press, 2019). For byproducts of the expanded computer industry, there is Jeffery R. Yost, *Making IT Work: A History of the Computer Services Industry* (Cambridge, Mass.: MIT Press, 2017); and Charles P. Bourne and Trudi Bellardo Hahn, *A History of Online Information Services, 1963–1976* (Cambridge, Mass.: MIT Press, 2003).

For more contemporary views, which I documented, see James W. Cortada, *EDP Costs and Charges: Finance, Budgets, and Cost Control in Data Processing* (Englewood Cliffs, N.J.: Prentice-Hall, 1980), *Managing DP Hardware: Capacity Planning, Cost Justification, Availability, and Energy Management* (Englewood Cliffs, N.J.: Prentice-Hall, 1983), *Strategic Data Processing: Considerations for Management* (Englewood Cliffs, N.J.: Prentice-Hall, 1984). But see also Leonard Rico, *The Advance Against Paperwork: Computers, Systems, and Personnel* (Ann Arbor: Graduate School of Business Administration, University of Michigan, 1967); R.A. Hirschheim, *Office Automation: A Social and Organizational Perspective* (New York: John Wiley & Sons, 1985); J.F. Rockart and D.W. DeLong, *Executive Support Systems: The Emergence of Top Management Computer Use* (Homewood, Ill.: Dow Jones-Irwin, 1988); P.G. W. Keen and M.S. Morton, *Decision Support Systems: An Organizational Perspective* (Reading, Mass.: Addison-Wesley, 1978).

Implications of more digital technologies, information can be explored in publications influential on management when they appeared. See, James M. Utterback, *Mastering the Dynamics of Innovation* (Boston, Mass.: Harvard

Business School Press, 1994); Clayton M. Christensen, *The Innovator's Dilemma: When New Technologies Cause Great Firms to Fail* (Boston, Mass.: Harvard Business School Press, 1997); Carl Shapiro and Hal R. Varian, *Information Rules: A Strategic Guide to the Network Economy* (Boston, Mass.: Harvard Business School Press, 1999); Thomas S. Wurster, *Blown to Bits: How the New Economics of Information Transforms Strategy* (Boston, Mass.: Harvard Business School Press, 2000). More tactically about how computers were being used, James W. Cortada, *Best Practices in Information Technology: How Corporations Get the Most Value from Exploiting Their Digital Investments* (Upper Saddle River, N.J.: Prentice Hall PTR, 1998) and *21st Century Business: Managing and Working in the New Digital Economy* (Upper Saddle River, N.J.: Financial Times/Prentice Hall, 2001); Thomas H. Davenport, *Process Innovation: Reengineering Work through Information Technology* (Boston, Mass.: Harvard Business Review Press, 1992); Michael Hammer, *Beyond Reengineering: How the Process-Centered Organization is Changing Our Work and Our Lives* (New York: Harper Business, 1996).

On supply chains and other collections of more recent types of information, useful introductions include Michael Hugos, *Essentials of Supply Chain Management*, 2nd ed. (Hoboken, N.J.: John Wiley & Sons, 2006): 133–168; Jack R. Meredith and Scott M. Shafer, *Operations and Supply Chain Management for MBAs*, 7th ed. (Hoboken, N.J.: John Wiley & Sons, 2019): 240–328; Victor Zarnowitz, *Business Cycles: Theory, History, Indicators, and Forecasting* (Chicago, Ill.: University of Chicago Press, 1996); Christopher M. Bishop, *Pattern Recognition and Machine Learning* (Berlin: Springer-Verlag, 2006) for a technical "how to" text; Cortada, *Information and the Modern Corporation*. Since much IT is being integrated together, such as the Internet of Things, AI, robotics, more Internet, and so forth, a useful composite review of current and immediately anticipated developments is Bob Tapscott, *Trivergence: Accelerating Innovation with AI, Blockchain, and the Internet of Things* (Hoboken, NJ: John Wiley & Sons, 2024).

About the Internet, I find many useful studies, but this one in particular stands out: Shane Greenstein, *How the Internet Became Commercial: Innovation, Privatization, and the Birth of a New Network* (Princeton, N.J.: Princeton University Press, 2015). On the role of children and young adults, a good place to start is with John Palfrey and Urs Gasser, *Born Digital: Understanding the First Generation of Digital Natives* (New York: Basic Books, 2008, 2016) and their *The Connected Parent: An Expert Guide to Parenting in a Digital World* (New York: Basic Books, 2020). Regarding the Cold War, see Paul N. Edwards, *The Closed World: Computers and the Politics of Discourse in Cold War America* (Cambridge, Mass.: MIT Press, 1996); Colin B. Burke, *America's Information Wars: The Untold Story of Information Systems in America's Conflicts and Politics from World War II to the Internet Age* (Lanham, Md.: Rowman & Littlefield, 2018). Also essential are Brian McCullough, *How the Internet Happened: From Netscape to the iPhone* (New York: W.W. Norton, 2018); Katie Hafner and Matthew Lyon, *Where Wizards Stay Up Late: The Origins of the Internet* (New York: Simon & Schuster, 1996); Janet Abbate, *Inventing the Internet* (Cambridge, Mass.: MIT Press, 1999); Barry Wellman and Caroline Haythornthwaite (eds.), *The Internet in Everyday Life* (Oxford: Blackwell, 2002); Lee Rainie and Barry Wellman, *Networked: The New Social Operating System* (Cambridge, Mass.: MIT Press, 2012); William Aspray and Barbara M. Hayes (eds.), *Everyday Information: The Evolution of Information Seeking in America* (Cambridge, Mass.: MIT Press, 2011). One should not overlook Mark Muro, Sifan Liu, Jacob Whiton, and Siddharth Kulkarni, *Digitization and the American Workforce* (Washington, D.C.: Brookings Institution, November 2017) for a recent study. Finally, for descriptions of what kind of information was required by hundreds of professions over the past 170 years, there is, for the United States, the U.S. Bureau of Labor Statistics, *Occupational Outlook Handbook*, published in many editions since the 1870s and in recent decades available online, https://www.bls.gov/ooh/.

LIVING IN THE INFORMATION AGE

A number of us have explored this issue. I have in James W. Cortada, *Living with Computers: The Digital World of Today and Tomorrow* (Cham, Switzerland: Springer, 2020), *How Societies Embrace Information Technology: Lessons for Management and the Rest of Us* (Hoboken, N.J.: John Wiley & Sons, 2009), and in *All the Facts: A History of Information in the United States since 1870* (New York: Oxford University Press, 2016). A useful book not to overlook is by Michael Buckland, *Information and Society* (Cambridge, Mass.: MIT Press, 2017). On the economics of a knowledge-centered society, a recent addition to the literature is by Joseph E. Stiglitz and Bruce C. Greenwald, *Creating a Learning Society: A New Approach to Growth, Development, and Social Progress* (New York: Columbia University Press, 2014), where they also make suggestions for policymakers. If one could only consult a single source, let it be Frank Webster, *Theories of the Information Society* 4th ed. (London: Routledge, 2014, first published 1996); and if only one history book, then Toni Weller, *Information History—An Introduction: Exploring an Emergent Field* (Oxford: Chandos Publishing, 2008). For a more intellectual discourse, see Mario Perez-Montoro, *The Phenomenon of Information: A Conceptual Approach to Information Flow* (Lanham, Md.: Scarecrow Press, 2007). For a detailed socio-technical analysis, one of the most widely respected students of the subject is Manuel Castells, and the crown jewel of his many publications is his trilogy, *The Rise of the Network Society, the Information Age: Economy, Society and Culture* (Oxford: Blackwell, 1996), *The Power of Identity, the Information Age: Economy, Society and Culture* (Oxford: Blackwell, 1997), and *The End of the Millennium, the Information Age: Economy, Society and Culture* (Oxford: Blackwell, 1997). Ronald E. Day's short book, *The Modern Invention of Information: Discourse, History, and Power* (Carbondale: Southern Illinois University Press, 2001) provides a useful window into modern thinking about information issues in modern society. It is quite thoughtful and not to be overlooked. Finally, a short thoughtful description of the role of expertise—academic, professional, and that of ordinary citizens—can be found in Harry Collins, *Are We All Scientific Experts Now?* (Cambridge: Polity, 2014).

There are many reminders that long before our time, earlier societies used information. For our project, I found the following relevant: Jürgen Habermas, *The Structural Transformation of the Public Sphere: An Inquiry a Category of Bourgeois Society* (Cambridge, Mass.: MIT Press, 1989); Daniel Headrick, *When Information Came of Age: Technologies of Knowledge in the Age of Reason and Revolution, 1700–1850* (New York: Oxford University Press, 2000); Anthony Giddens, *Social Theory and Modern Sociology* (Cambridge: Polity Press, 1987); Edward Higgs, *The Information State in England: The Central Collection of Information on Citizens Since 1500* (Basingstoke: Palgrave Macmillan, 2004), but do not overlook his other book, *Life, Death and Statistics: Civil Registration, Censuses and the Work of the General Register Office, 1836–1952* (Hatfield: University of Hertfordshire Press, 2004); and Jon Agar, *The Government Machine: A Revolutionary History of the Computer* (Cambridge, Mass.: MIT Press, 2003). Do not overlook the important study of how information has defined who we are through the forms we fill out, the identification cards we use, and the financial and tax records kept on us, all of which are well described by Colin Koopman, *How We Became Our Data: A Genealogy of the Informational Person* (Chicago, Ill.: University of Chicago Press, 2019).

Alvin Toffler's key works include *Future Shock* (New York: Random House, 1970) and *The Third Wave* (New York: William Morrow, 1980). I am not aware of a biography about him. On the French experience, we have Alain Minc's memoirs, *Voyage ou Centre du "Système"* (Paris: Bernard Grosset, 2019) and an accessible edition of the French study is available: Simon Nora and Alain Minc, *The Computerization of Society: A Report to the President of France* (Cambridge, Mass.: MIT Press, 1980). The standard work on the Minitel is now Julien Mailland and Kevin Driscoll, *Minitel: Welcome to the Internet* (Cambridge, Mass.: MIT Press, 2017).

THE RISE OF BIG DATA

Big Data is a relatively new form of information technology, so much of the literature on it focuses on explaining what it is or consists of technical publications on how to implement and use it. So much of the information about it has to be gleaned from technical journals and the trade press. The ones used in chapter 4 can be found in its endnotes. But a few books are beginning to appear. A useful entry point into the subject is David Weinberger, *Too Big to Know: Rethinking Knowledge Now That the Facts Aren't the Facts, Experts Are Everywhere, and the Smartest Person in the Room Is the Room* (New York: Basic Books, 2012), which is not a technical treatise. It might also be useful to consider how software is changing as Big Data rises, so for that consult D.M. Berry, *The Philosophy of Software: Code and Mediation in the Digital Age* (London: Palgrave Macmillan, 2011) and a surprisingly good read, Daniel E. Atkins et al., *Revolutionizing Science and Engineering Through Cyberinfrastructure: Report of the National Science Foundation Blue-Ribbon Panel on Cyberinfrastructure* (Washington D.C.: National Science Foundation, January 2003). For insight on how scientists are reacting to Big Data, an excellent starting point is the short but very approachable report by David Bollier, *The Promise and Peril of Big Data* (Washington, D.C.: Aspen Institute, 2010). For a criticism of how firms and governments use Big Data, the key publication is Shoshana Zuboff and James Maxmin, *The Age of Surveillance Capitalism: The Fight for a Human Future at the New Frontier of Power* (New York: PublicAffairs, 2019). Two books by Pablo J. Boczkowski are essential sources on the role of such data sets in society: Pablo J. Boczkowski, *Abundance: On the Experiences of Living in a World Information Plenty* (New York: Oxford University Press, 2021), and with Eugenia Mitchelstein, *The Digital Environment: How We Live, Learn, Work, and Play Now* (Cambridge, Mass.: MIT Press, 2021). For a combination of personal memoir and academic discourse written by an information science expert, see Melanie Feinberg, *Everyday Adventures with Unruly Data* (Cambridge, Mass.: MIT Press, 2022).

For more technical explanations, there are now numerous texts. I used O'Reilly Media, *Big Data Now* (Sebastopol, Cal.: O'Reilly Media, 2011). One cannot stray too far from how the mind works when dealing with such topics as Big Data. Nikolas Rose and Joelle M. Abi-Rached, *Neuro: The New Brain Sciences and the Management of the Mind* (Princeton, N.J.: Princeton University Press, 2013). There are a series of publications that relate to Big Data that can help situate this new development into current technological and biological realities: Robert R. Korfhage, *Information Storage and Retrieval* (New York: John Wiley & Sons, 1997); Richard K. Belew, *Finding Out About: A Cognitive Perspective on Search Engine Technology and the W.W.W.* (Cambridge: Cambridge University Press, 2000); Michael Lesk, *Understanding Digital Libraries*, 2nd ed. (San Francisco, Cal.: Morgan Kaufmann, 2005). For three texts on metadata, see Richard Gartner, *Metadata: Shaping Knowledge from Antiquity to the Semantic Web* (Berlin: Springer-Verlag, 2016); Marcia Zeng and Jian Qin, *Metadata* (New York: ALA Neal-Schuman, 2016); Steven J. Miller, *Metadata for Digital Collections* (New York: ALA Neal-Schuman, 2020). For a blend of discussions concerning the role of technology and user behaviors, we now have the most useful volume by Mark Shepard, *There Are No Facts: Attentive Algorithms, Extractive Data Practices, and the Quantification of Everyday Life* (Cambridge, Mass.: MIT Press, 2022).

FAKE FACTS, FAKE NEWS

Recent events have supercharged interest in such topics as "fake facts," "alternative facts," and the more mundane issues of lies, rumors, misinformation, conspiracies, and gossip. Each existed for centuries and, when studied by scholars, was often done in a piecemeal fashion. Political scientists have done the most to understand the presentation and manipulation of truth and facts for political purposes, such as how to win elections. In the past two decades, experts on media and advertising have contributed to a more theoretically based understanding of the role of all manner of facts, both truthful and not. Historians are just beginning to weigh in on the matter, too. William Aspray and I, both trained as historians, have recently weighed in with two studies: James W. Cortada and William Aspray, *Fake News Nation: The Long History of Lies and Misinterpretations in America* (Lanham, MD: Rowman & Littlefield, 2019) and William Aspray and James W. Cortada, *From Urban Legends to Political Fact-Checking: Online Scrutiny in America, 1990–2015* (Cham, Switzerland: Springer, 2019). A group of media experts have also provided varying perspectives and brief case studies on related issues, focusing largely on recent examples: Melissa Zimdars and Kembrew McLeod (eds.), *Fake News: Understanding Media and Misinformation in the Digital Age* (Cambridge, Mass.: MIT Press, 2020). We did not discuss spying, espionage, or the work of governments to misinform each other, but it has a rich history. For a well-informed account of Twentieth Century activities, see Thomas Rid, *Active Measures: The Secret History of Disinformation and Political Warfare* (New York: Farrar, Straus and Giroux, 2020).

For the language of the politician, a useful place to begin understanding political rhetoric, see Murray Edelman, *Political Language: Words That Succeed and Politics that Fail* (New York: Academic Press, 1977). This political scientist wrote a raft of books on American political behavior that are core to understanding the role of information in national elections; one of the most useful of which is *Politics As Symbolic Action: Mass Arousal and Quiescence* (Chicago, ILL: Markham, 1971). But see also his *The Symbolic Uses of Politics* (Urbana, ILL.: University of Illinois Press, 1985). Do not overlook Dan F. Hahn, *Political Communication: Rhetoric, Government, and Citizens* (State College, PA.: Strata Publishing, 1998), or and Bruce Bimber, *Information and American Democracy: Technology in the Evolution of Political Power* (Cambridge: Cambridge University Press, 2003). All have rich bibliographies. Presidential assassinations are about conspiracies and have been studied extensively. On Lincoln's, the most useful is William Hanchett, *The Lincoln Murder Conspiracies* (Urbana, ILL.: University of Illinois Press, 1983). On mythologizing him, see Gabor Borritt, *The Lincoln Enigma: The Changing Faces of an American Icon* (New York: Oxford University Press, 2001); Philip B. Kunhardt III, Peter W. Kunhardt, and Peter W. Kunhardt, Jr., *Looking for Lincoln: The Making of an American Icon* (New York: Alfred A. Knopf, 2008); and Merrill D. Peterson, *Lincoln in American Memory* (New York: Oxford University Press, 1994). On Kennedy's, there is Jim Marrs, *Crossfire: The Plot that Killed Kennedy* (New York: Basic Books, 2013). The best-researched study, however, is by Philip Shenon, *A Cruel and Shocking Act: The Secret History of the Kennedy Assassination* (New York: Henry Holt, 2013).

The use of information by companies and industries has been explored. For a good introduction to a number of these, consult Ari Rabin-Havt and Media Matters for America, *Lies, Incorporated: The World of Post-Truth Politics* (New York: Anchor Books, 2016). The tobacco industry's behavior can be studied thoroughly with Richard Kluger, *Ashes to Ashes: America's Hundred-Year Cigarette War, the Public Health, and the Unabashed Triumph of Philip Morris* (New York: Knopf, 1996); Philip J. Hilts, *Smoke Screen: The Truth Behind the Tobacco Industry Cover-Up* (Boston, MA: Addison-Wesley, 1996); David Kessler, *A Question of Intent: A Great American Battle with a Deadly Industry* (New York: Public Affairs, 2001); Allan Brandt, *The Cigarette Century: The Rise, Fall and Deadly Persistence of the Product That Defined America* (New York: Basic Books, 2007); and Naomi Oreskes and Erik M. Conway, *Merchants of Doubt: How a Handful of Scientists Obscured the Truth on Issues from Tobacco Smoke to Global Warming* (New York: Bloomsbury, 2010). This last book is the most authoritative study currently available based on a massive cache of industry documents.

On global warming, consult James Inhofe, *The Greatest Hoax: How the Global Warming Conspiracy Threatens Your Future* (Washington, D.C.: WND Books, 2012). For the alternative view, see National Research Council, *Limiting the Magnitude of Future Climate Change* (Washington, D.C.: The National Academic Press, 2010). Historical perspective combined with the use of a growing body of scientific evidence of the Earth's evolving climate over the millennia can be found in S. Fred Singer and Dennis T. Avery, *Unstoppable Global Warming: Every 1,500 Years* (Lantham, MD.: Rowman & Littlefield, 2007). The most widely available discussion of the issues, demonstrating alternative uses of similar data, rhetoric, and opinions can be found in Al Gore, *An Inconvenient Truth: The Planetary Emergency of Global Warming and What We Can Do About It* (Emmaus, PA: Rodale, 2006) and a video he produced with the same title (New York: Melcher Media, 2006). For a more academic discussion along similar lines as Gore's, see National Research Council, *Climate Change: Evidence, Impacts, and Choices* (Washington, D.C.: National Research Council, 2011). Finally, on the manipulation of information, see David Michaels, *Doubt Is Their Product: How Industry's Assault on Science Threatens Your Health* (New York: Oxford University Press, 2008).

On business uses of information, particularly public relations and advertising, begin with Stuart Ewen, *PR! A Social History of Spin* (New York: Basic Books, 1996) and Roland Marchand, *Creating the Corporate Soul: The Rise of Public Relations and Corporate Imagery in American Big Business* (Berkeley, CA.: University of California Press, 1998), but also Fred Koenig, *Rumors in the Marketplace: The Social Psychology of Commercial Hearsay* (New York: Praeger, 1985), and folklorist Gary Alan Fine, *Manufacturing Tales: Sex and Money in Contemporary Legends* (Knoxville, TN.: University of Tennessee Press, 1992). For a useful introduction to the history of advertising, see Daniel Pope, *The Making of Modern Advertising* (New York: Basic Books, 1983). Patent medicines have an extensive literature, but the authority is James Harvey Young, *The Toadstool Millionaires: A Social History of Patent Medicines in America Before Federal Regulation* (Princeton, NJ.: Princeton University Press, 1961), *The Medical Messiahs: A Social History of Health Quackery in Twentieth-Century America* (Princeton, NJ.: Princeton University Press, 1967), and *American Health Quackery* (Princeton, NJ.: Princeton University Press, 1992). For a study of nutrition and wellness information, see the early study by Victor Herbert, *Nutrition Cultism: Facts and Fictions* (Philadelphia, PA.: George F. Stickley, 1980).

An excellent place to begin regarding the Internet and facts is Laura Gurak, *Cyberliteracy: Navigating the Internet with Awareness* (New Haven, CT: Yale University Press, 2001), followed by Faithe Wempen, *Digital Literacy for Dummies* (Hoboken, NJ: John Wiley & Sons, 2015), Nicole A. Cooke, *Fake News and Alternative Facts: Information Literacy in a Post-Truth Era* (Chicago, IL: ALA Editions, 2018), and Daniel J. Levitin, *A Field Guide to Lies: Critical Thinking in the Information Age* (New York: Dutton, 2016). On metaliteracy, a key source is Thomas P. Mackey, *Metaliteracy: Reinventing*

Information Literacy to Empower Learners (Chicago, IL: American Library Association, 2014).

Finally, about how the human mind deals with information, the role of bias, and errors in historical context, there is the highly readable authoritative account by Daniel Kahneman, Oliver Sibony, and Cass R. Sunstein, *Noise: A Flaw in Human Judgment* (New York: Little, Brown Spark, 2021).

LOOKING AT INFORMATION TODAY

In addition to earlier studies about how the brain works, I would recommend several that are more application oriented. These include Paul B. Armstrong, *Stories and the Brain: The Neuroscience of Narrative* (Baltimore Md.: Johns Hopkins University Press, 2020); Bob Garrett and Gerald Hough, *Brain & Behavior: An Introduction to Behavioral Neuroscience*, 5th ed. (Thousand Oaks, Cal.: SAGE, 2018); and on the role of bias, Pamela Fuller, Mark Murphy, and Anne Chow, *The Leader's Guide to Unconscious Bias: How to Reframe Bias: Cultivate Connection, and Create High-Performance Teams* (New York: Simon & Schuster, 2020); Erin Beeghly and Alex Madva (eds.), *An Introduction to Implicit Bias: Knowledge, Justice, and the Social Mind* (London: Routledge, 2020); Howard J. Ross, *Everyday Bias: Identifying and Navigating Unconscious Judgments in Our Daily Lives* (Lanham, Md.: Rowman & Littlefield, 2020) and especially his bibliography, pp. 173–180. Because of so much misinformation existing, we need to improve our ability to scrutinize what we see. For that, I suggest Jennifer Kavanaugh and Michael D. Rich, *Truth Decay: An Initial Exploration of the Diminishing Role of Facts and Analysis in American Public Life* (Santa Monica, Cal.: RAND, 2018); Jason Altmire, *Dead Center: How Political Polarization Divided America and What to Do About it* (Mechanicsburg, Penn.: Sanbury Press, 2017). In collaboration with William Aspray, we have provided prescriptive insights, William Aspray and James W. Cortada, *From Urban Legends to Political Fact-Checking: Online Scrutiny in America, 1990–2015* (Cham, Switzerland: Springer, 2019). More broadly, I have expanded on some of these themes in James W. Cortada, *Living with Computers: The Digital World of Today and Tomorrow* (Cham, Switzerland: Springer, 2020).

On the controversial matter of IQ tests and intelligence, I relied extensively on James R. Flynn, *What Is Intelligence? Beyond the Flynn Effect* (Cambridge: Cambridge University Press, 2007), who studied the structure and uses of IQ tests. For the argument that humans possess multiple intelligences, the standard work is Howard Gardner, *Multiple Intelligences* (New York: Basic Books, 1993, 2006) and *Frames of Mind: The Theory of Multiple Intelligences* (New York: Basic Books, 1983, 2004, 2011). On how trees communicate, the most widely cited study is Peter Wohlleben, *The Hidden Life of Trees: What They Feel, How They Communicate* (Vancouver, Can.: Greystone Books, 2015).

On the power of group think, generalists, and collaboration, begin with Richard H. Thaler and Cass R. Sunstein, *Nudge: The Final Edition* (New York: Penguin, 2009, 2nd ed. 2021), which, like Kahneman's study, discusses how to use biases to alter behavior and thinking.

On using information in historical perspective, see David Epstein, *Range: Why Generalists Triumph in a Specialized World* (New York: Riverhead Books/Penguin, 2019): 55–214. I commented on this phenomenon too in James W. Cortada, *History Hunting: A Guide for Fellow Adventurers* (Armonk, NY: M.E. Sharpe, 2012): 223–247, but see also Richard E. Neustadt and Ernest R. May, *Thinking in Time: The Uses of History for Decision Makers* (New York: Free Press, 1986); Gordon C. Wood, *The Purpose of the Past: Reflections on the Uses of History* (New York: Penguin, 2009). For a highly approachable perspective by two psychologists, see Steven Sloman and Philip Fernbach, *The Knowledge Illusion: Why We Never Think Alone* (New York: Riverhead/Penguin, 2017).

Index

9/11 attacks, 123–25

academic discipline, 22, 24, 61, 75, 94, 161; collaboration, 54–55, 95, 100–101; -specific definitions, 4–6
accounting, 40
accuracy, 15
adaptive cycle, 19–20, 79
advertising: false, 115; Watson, 172
agency, 13
agriculture, workforce, 11–12
algorithms, 38, 122, 161
Al-Khalili, Jim, 163
Amazon, 57, 63, 90, 103, 146; data collection, 89
Amnesty International, 104
analytics, 48; predictive, 90, 95, 98
Anderson, Chris, 106–7
anthropology, 16
Apple, 41, 103–4
applications, 46; accounting, 40; computer, 38; industry-specific, 43–44; inventory management, 41; spreadsheet, 50; supply chain, 51
appropriation, 19
artificial intelligence (AI), 7–8, 23–24, 29, 31–32, 54, 63, 98, 112, 136, 157, 167; ChatGPT, 173; decision making, 175; generative, 158, 175; historical perspective, 169–70, 174–75; hype, 170–72; impact on the workplace, 174; machine learning, 92; medical/healthcare applications, 170–71, 174; pattern recognition, 162–63; regulation, 168, 174; Sora, 175–76; supply chain, 50–51; training, 175; Watson, 170–72
Aspray, William, 120, 124, 126
AT&T, 44
augmentation, 44
automation, 36–37, 51, 80, 90, 157; banking, 44; islands of, 53; magnetic ink character recognition, 40; manufacturing, 39–40; point-of-sale systems, 44; robotics, 177
automobile manufacturing, 76; supply chain, 48–49
availability bias, 140

Baader-Meinhof phenomenon, 9–10
banking, 40, 49; automation, 44
barcodes, 50
belief, 113, 128; religious, 141–42
Bell, Daniel, 75, 81
Bezos, Jeff, 146
bias, 20, 102, 138–40, 160; availability, 140; cognitive, 9–10; competency,

233

140–41; familiarity, 140; perseverance, 141
Big Data, 14, 24, 28, 83–84, 161, 175; versus Big Data, 108; Cambridge Analytica data breach, 86; collection, 102–3; commercial use, 99; context, 101–2; data discovery, 93; data mining, 92, 95–97; definitions, 89; disease surveillance, 86–87; ethics, 102–3; hubris, 87–88, 91, 94; implications for theory-based science, 105–7; insight, 92–93; opacity, 97; privacy issues, 102–3; research impacts, 101; research methodology, 96; in retail, 92; retrieval, 100; scientific uses, 92; sources, 100; statistical analysis, 90; storage, 89; subjectivity, 96; value, 90; variability, 90; variety, 89–90; veracity, 90, 109; volume, 89, 95
Bigelow, Julian, 178
bioinformatics (BI), 24
biology, 3–5, 7–8, 13, 22–26, 29, 137, 144–45. *See also* ecosystem
blogs, 131
Bluetooth, 65
Boczkowski, Pablo J., 100–101
books, 9–10, 12, 20, 30; reference, 70. *See also* publications
Borgman, Christine L., 100
bots, 128
Boyd, Danah, 101
brain science, 9–10, 23–25, 90, 137–38; bias, 140. *See also* bias; neurons, 147; situational awareness, 139
Breznitz, Dan, 102–3
Bridle, James, 13, 172
Brookings Institution, 59, 141
Buckland, Michael, 69
burden of knowledge, 2
Burke, Peter, 2, 27
business intelligence, 48

Cambridge Analytica data breach, 86
cancel culture, 142
card catalog, 71

Cash, Johnny: "A Boy Named Sue", 67
Castells, Manuel, 75
certainty, 12–13, 90
change, 79
ChatGPT, 173
chemistry, 23–24
chief executive officer (CEO), 54
chief financial officer, 38
chief information officer (CIO), 53
China, 77
chronology, 67–68
climate change, 73, 87, 120, 122
cloud computing, 54
cognitive bias, 9–10
cohabitation, 146
Cold War, 119
collaboration, 54–55, 95, 142–43, 159; cross-disciplinary, 100–101
collective intelligence, 17, 158–59
communication, 15–16; storytelling, 26–27. *See also* e-mail; infrastructure; network(ing)
community, 13, 16, 22, 142; collective intelligence, 17
competency bias, 140–41
complexity, 5–6, 24
computer/computing, 35, 64, 113, 144; applications, 38; cloud, 54; commercial use, 39–40; data centers, 38; data processing, 40; dissemination of fake facts, 121–22; early uses, 38–41; experimental, 38; first and second wave adopters, 43; French study, 80; "heart transplant" 41; IBM System 360, 41–42; input/output, 38; inventory management, 41; laptop, 46–47; literacy, 130; mainframe, 38; in manufacturing, 40–41; mini, 46–47; operating system, 149; personal, 45–51; processes, 150–51; programming, 8; science, 7–8, 24; screen time, 136; system, 38, 150–51; von Neumann architecture, 150; "what if" analysis, 38–39. *See also* e-mail
conceptual thinking, 149–50

conspiracy theory, 112, 116, 118, 142; 9/11, 123–25
consumption of information, 26
contact tracing, 65, 109
content, 15
context, 5, 15–16, 101–2, 137–39, 160–63
contracts, 58
control, information, 107–8
convergence, 54
Cooke, Nicole A., 130
copyright, 70, 97–98
correlation, 87, 106–7
corruption, 127
Cortada, James, 124; *All the Facts*, 3, 12, 74, 82, 144, 163; *Birth of Modern Facts*, 3, 18, 35, 67, 74, 82, 144, 148, 163
COVID-19 49, 51, 96, 108–9, 112; contact tracing, 65, 109; "flattening the curve" strategy, 85–86; hotspots, 65, 87; lockdowns, 136–37; news, 65; vaccine, 116
Crawford, Kate, 101
critical thinking, 132
crowdsourcing, 159
cryptography, 57–58
cyber libel, 127

data, 4–5, 23, 25–26; breach, 86; cartel, 98; collection, 7, 88–89, 91, 102–5; discovery, 93; versus information, 90–91; meta, 108; mining, 92, 95–97; processing, 28, 53; qualitative, 94–96; quantitative, 94–96; raw, 96; storage, 91; value, 89; velocity, 89; warehouse, 92. *See also* Big Data
database, 20, 28, 47, 71, 87, 102; medical, 23; relational, 60, 88
data centers, 38, 41, 47, 52; management, 45
Day, Ronald E., 75
debunking, 117
decision making, 157; AI, 175; exogenous events, 97; heuristics, 141, 160, 174; objectivity, 142; situational awareness, 139. *See also* bias; brain science
deep learning, 162
definitions, 27
DeLong, J. Bradford, 33
Deming, W. Edwards, 44, 48, 54
Denning, Peter J., 91
d'Estaing, Valéry Giscard, 80
DeVito, Mike, 129
Dewey, Melvil, 71
diffusion, 37, 43, 46–47, 56; digital device, 57
digital devices, 57, 75–76, 90
digital plumbing, 51–53; convergence, 54
digital signature, 57–58
digitization, 59
Dillon, Andrew, 5
discipline. *See* academic discipline
disease surveillance, 86–87
disinformation, 117
DNA, 7, 15, 24
documents, 69, 92; copyright, 97–98; digital, 57–58; reference books and manuals, 70
Dyson, George, 159, 178

Eckert-Mauchly Computer Company, 38
ecology, information, 145
economics: GDP, 12, 52, 72; standard of living, 73
ecosystem, 16, 22–23, 25; adaptive cycle, 19–20; context, 160–63; information, 5, 10, 13, 15, 17, 30, 51, 61, 137, 139, 143–48, 151
Eddington, Arthur, 14
edge, 14–15
education: formal, 11, 131, 155–56; IQ and, 156; job requirements, 71; Luria on, 21–22; policy, 73
electricity, 24, 28, 49, 52
electrified information, 6
Electronic Signatures in Global and National Commerce Act (US, 2000), 58
Ellul, Jacques, 149

Elsevier, 98
Elvery, Joel A., 37
e-mail, 43, 45, 47, 55–56, 72–73
embodied cognition, 177–78
emotional intelligence (EI), 158
endowment, 74
ENIAC, 38
Ensmenger, Nathan L., 8
epistemology, 101
espionage, 112–13
ethics, 101; Big Data, 102–3
ethnography, 16
European Union, 24, 63, 103–4
experimentation, 7
experts/expertise, 9, 112, 117, 128, 136; Big Data, 94–95, 97–98; credentials, 118, 141, 164–65

FAANG (Facebook, Amazon, Apple, Netflix, and Alphabet), 102–3
Facebook, 73, 87, 90, 103, 129, 153; Cambridge Analytica data breach, 86; data collection, 89; dissemination of false information, 126–27; "Like" button, 104
fact(s), 2–5, 7–8, 10, 13, 20, 26, 30, 71, 88, 93; checking, 120, 122, 124–26, 128, 130–31; creating, 93; fake. *See* fake facts; misrepresentation, 113–14
fake facts, 83–84, 93, 111–12, 119–20, 170; conspiracy theory, 116; debunking, 117; disinformation, 117; on the Internet, 121–27; misinformation, 117; misleading information, 113–14; misrepresentation of the truth, 113–15; political untruths, 114–15; rhetoric, 118; role of computing in dissemination, 121–22; rumor, 116; urban legend, 123–24
fake news, 63, 116–17, 120, 122, 124, 129
false advertising, 115
familiarity bias, 140
feedback loop, 14
Feinberg, Melanie, 90

"flattening the curve" strategy, 85–86
flexible manufacturing, 44
Flynn, James R., 155–58
formal education, 11, 131, 155–56
framing, 140–41
France, 80; *L'informatisation de la société*, 80–81; *Minitel*, 81–82
Frankfurt, Harry G., 114
freedom of expression, 127
frequency illusion, 9–10
Friedman, Benjamin M., 8, 22
de Gaulle, Charles, 80

GDP, 12, 52, 72–73, 94
generalist, 161
generative AI, 158, 175
genetics, 23–24
genomics, 24
Giddens, Anthony, 77
Google, 65, 87, 103, 105, 108, 161, 173; privacy policy, 105. *See also* privacy
Gordon, Robert J., 11
Graff, Harvey J., 176
Griffel, David, 108
group intelligence, 158–59

Habermas, Jürgen, 77
Haidt, Jonathan, 25, 147
hard sciences, 1–2, 4, 26; mathematics, 21–22
Headrick, Daniel, 77–78
healthcare, 36–37; artificial intelligence (AI), 170–71; Big Data, 99; false advertising, 115; smart thermometers, 86
heuristics, 141, 160
Higgs, Edward, 77
Hinton, Geoffrey, 167
historians, 161, 164–65; period names, 66–70; quantitative versus qualitative, 94–95
hotspots, COVID-19, 65, 87
hubris, 87–88, 91, 94, 154, 171–73
humanoid robots, 177–78
Huntington, Samuel P., 76–77

Index

IBM, 38, 45, 81, 83, 132; Lotus Notes, 52; System 360, 41–42; Watson, 170–72
identity, political, 141–42
ignorance, 2
imaging technology, 92
inclination, 10
Industrial Revolution, 67, 155
information, 1, 32–33, 65, 164; accuracy, 15; appropriation, 19; bias, 9–10, 20, 102, 138–39; cartel, 98; complexity, 6, 24; consumption, 26; content, 15; context, 15–16, 160–63; control, 107–8; convergence, 54; copyright, 97–98; creating, 93; creation, 10; versus data, 90–91; definitions, 4–10, 13; diffusion, 37, 43, 46–47, 56; digitization, 57–59; discipline-specific definitions, 5–6; "disorder", 117; ecology, 145; ecosystem, 5, 10, 13, 15, 17, 19, 22, 25, 30, 51, 61, 137, 139, 143–48, 151; electrified, 6, 28–29; growth rate, 2, 6; illiteracy, 127–29; infrastructure, 13, 15, 18, 30, 51–52; innovation and, 161–62; intellectual property, 74; inventory, 41; language, 26; literacy, 129–33; management, 100; metaliteracy, 132–33; mis- 117, 119–20; misleading, 113–14; modeling, 60–61; narrative intelligence, 4; "on demand", 43, 60; organized, 5, 12; overload, 74–75, 78–79; packets, 55–56; pattern recognition, 51; "pollution", 117; precision, 21–22; product sales, 72; providers, 98; public administration, 77; real time, 50; scaling, 119–20; scrutiny, 127–28; security, 112–13; sharing, 144; society, 77–78; specialization, 2–3, 135–36, 144; spreadsheets, 50; storage, 7–8; as a tool, 8; usefulness, 19; weaponized, 118–19. See also Big Data

Information Age, 64–65, 82–83; arguments for why we don't live in, 76–78; arguments for why we live in, 73–76; economic and social impact, 71–73; institutional responses, 70–71; naming, 66–70; quantity of information created and used, 71
information and computing technology (ICT), 81
The Information Society, 75
information technology (IT), 35; language, 148–49
informed supply chain, 48
infrastructure, 15–16, 18; content, 15; digital plumbing analogy, 51–54; information, 13, 30; supply chain, 48–49
innovation, 11, 35–36, 68, 74, 161–62
insight, 92–93
intellectual property, 74
intelligence, 23, 154–55, 160; business, 48; collective, 17, 158–59; embodied cognition, 177–78; emotional, 158; narrative, 4
Internet, 4–5, 8, 12, 14, 18, 28–29, 47, 53, 59–60, 68, 113, 120, 144, 170; bots, 128; dissemination of misinformation and fake facts, 122–25, 128–29; fake websites, 128; information literacy, 129–33; packets, 55–56; Semantic Web, 30–32; work from home, 52; worldwide usage, 72. See also social media
Internet of Things (IoT), 90, 99
inventory management, 41; just-in-time approach, 50; RFID tags, 47
IQ test, 154–55; cohort comparisons, 156; scoring, 156–57
Islam, 77
islands of automation, 53

Jackson, Andrew, 118
journals, 75; copyright, 97–98; "predatory", 111
just-in-time approach, 50

Kennedy, John F., 118
keyboarding, 18
Killen, Andreas, 23
Kinsa Health, 86
knowledge, 4, 7, 18; burden of, 2; collective, 142–43; expertise, 9; scientific, 10; work, 36, 72
Koller, Julia, 129–30
Kuhn, Thomas, 6, 19

labor, 11–12, 19; productivity, 39–40. *See also* academic discipline; productivity; work; workforce
Laney, Doug, 89, 91
language, 11, 13, 18, 25; of information, 26; information technology, 148–49; translation, 92
laptops, 46–47, 65
Large Hadron Collider (LHC), 92
Large Synoptic Survey Telescope (LSST), 92
Le Point, 80
Lévi-Strauss, Claude, 16
library(ies), 72, 112, 146; university, 70–71
"Like" button, 104
Lincoln, Abraham, 118
L'informatisation de la société, 80–81
literacy, 10–11, 18, 72, 79, 82; computer, 130; information, 129–33; meta, 132–33
Longino, Helen E., 147
Luria, Alexander, 155–56; on education, 21–22

machine learning, 92, 172
Machlup, Fritz, 72
magnetic ink character recognition, 40
mainframe, 38, 45, 68; IBM System 360, 41–42
management, 35–36; data center, 45; data processing, 53; information, 100; supply chain, 51
Management Information System (MIS), 53

manufacturing: automation, 37, 39–40; computers, 40–41; flexible, 44; productivity, 36; robotics, 177
marketing, Watson, 171–72
massive computing, 24
mathematics, 23; precision, 21–22
Maxim, James: *The Age of Surveillance Capitalism: The Fight for a Human Future at the New Frontier of Power*, 104–5
McIntosh, Stuart, 108
medicine/medical: false advertising, 115; misinformation, 119–20
metadata, 108
metaliteracy, 132–33
methodology, 102; Big Data research, 96
microscope, 23
Microsoft, 57, 173
Minc, Alain, 80, 82
mini computer, 46–47
Minitel, 81
Miodownik, Mark, 30
misinformation, 112, 117, 119–20
misleading information, 113–14
misrepresentation, 113–15
modeling, 60–61, 153; conceptual, 149–50; language, 92
Moore's Law, 57
Mossberger, Karen, 129
Muratov, Dmitry, 127

naming historical periods, 66–70, 83
Nardi, Bonnie A., 147
narrative intelligence, 4
natural language processing, 162
nervous system, 14, 24–25
Netflix, 65
network(ing), 16, 19, 30, 55, 143; digital plumbing analogy, 51–54; edge, 14–15; node, 14–15; social, 13; telephone, 35, 44
von Neumann, John, 149–50
neurons, 147
neuroscience, 138
New Deal, 66

New York Times, 2, 79, 105, 109, 128
news, 142; aggregators, 122; COVID-19, 65–66; fact checking, 125–26, 128; fake, 63, 116–17, 120, 124, 129
Nixon, Richard, 119
node, 14–15
Nora, Simon, 80, 82
norms, 10, 13, 15
nudge, 160

objectivity, 142, 147
observation, 93, 113, 142
occupational groups, 36
O'Day, Vicki L., 147
office administration, 36
on-the-job training, 8
opacity, Big Data, 97
operating system, 149
Operations Research (OR), 176
opinion, 26–27; expert, 118
organized information, 5, 12
Otlet, Paul, 83–84

Page, Larry, 108
pareidol, 124
patents, 11, 27
pattern recognition, 51, 162–63
period names, 66–70
perseverance bias, 141
personal computers (PCs), 45, 56–57, 65; dissemination of fake facts, 121–22
personal responsibility, 131–32
petabyte, 92
Philippines, 126–27
philosophy, 4, 114. *See also* truth
platform, 60, 151, 153
point-of-sale systems, 44
Polanyi, Michael, "Republic of Science", 16
policy: education, 73; privacy, 105
politics/political, 27, 66, 132–33; conspiracy theory, 118, 125; fact checking, 125–26, 128; fake news, 63, 117–18; identity, 141–42; rhetoric, 118; scandal, 116–17; tribalism, 76–77, 141–42; weaponized information, 119
precision, 21–22
predictive analytics, 90, 95, 98
primary sources, Big Data, 100
privacy, 64–65, 70, 98, 102; Cambridge Analytica data breach, 86; data collection, 103–5; policy, 105
processes, 48, 54, 148–49; computer, 150–51; informed supply chain, 48
production: automation, 36–37; workers, 36. *See also* manufacturing
productivity, 11, 19, 36, 39–40, 73; AI and, 174; turnpike effect, 29
professions, 6; educational requirements, 71; IQ scores, 157–58
programming, 8
progress, 5–6, 20
propaganda, 115
psychology, 138, 154–56, 158; social, 25
public administration, 77
publications, 10–12; brain science, 138; copyright, 97–98; journal, 75; scientific, 23–24
public health, disease surveillance, 86–87

qualitative data, 94–96
quantitative data, 94–96

Radio Frequency Identification (RFID) tags, 47, 90
Rappler, 127
Rauch, Jonathan, 141
raw data, 96
reading, 11
real time, 50
regulations, 70, 98, 115, 132, 153, 168; artificial intelligence (AI), 168, 174
relational database, 60, 88
Renn, Jürgen, 107–8
Republic of Science, 16
research, 98, 161; Big Data impact on, 101; brain science, 138; and

development, 74; methodology, 96, 102; subjectivity, 96; team-based, 159
Ressa, Maria Angelita, 127
retail: Big Data, 92; point-of-sale systems, 44
revenge editing, 127
rhetoric, 118
Rid, Thomas, 112–13
robotics, 169, 176; humanoid, 177–78
"rolling apocalypse", 20
Rometty, Virginia, 171
rules, 7, 17; decision making, 174–75. *See also* heuristics
rumor, 116, 123–25, 142. *See also* conspiracy theory
Russia, 111–13

sales, information products, 72
scaling, information, 119–20
scandal, 116–17
Scharf, Caleb, 164
Schatzberg, Eric, 27, 30
Schiller, Herbert, 75, 77
Schmidt, Eric, 104
Schumpeter, Joseph, 74
science, 6, 14, 27; brain. *See* brain science; communications, 26–27; computer, 7–8, 24; misleading information, 113–14; objectivity, 147; publications, 23–24; Republic of, 16; theory-based, 105–7
scientific method, 7, 139, 147, 163
screen time, 136
scrutiny, 127–28
secondary sources, Big Data, 100
Second Industrial Revolution, 11
Seely Brown, John, 107
Semantic Web, 30–32
sensors, 40–41, 47, 55, 60, 87, 90–91, 108, 146–48, 163; supply chain, 49–50
servers, 55
Shannon, Claude E., 3, 6, 14, 49
signals, 14, 49, 145
situational awareness, 139

skills, 18; Big Data, 100; critical thinking, 132
slander, 114–15
smartphones, 56–57, 64, 71; contact tracing, 65, 109
smart speakers, 57
smart thermometers, 86, 96
Smith, Adam, 8
Snopes.com, 124–26
social media, 6, 63, 65, 117, 120, 142; fake news, 122; "likes", 123; privacy issues, 103–5
social network analysis, 13–14
social psychology, 25
social science, 13
sociology, 13
software, 24, 54, 87, 145, 148–49, 151, 154; AI, 173–74; generative, 173; industry-specific, 43–44
Solow, Robert, 73
Sora, 175–76
sources, Big Data, 100
specialization, 2–3, 8, 10, 70, 135–36, 142–44, 161–62, 176
spillover, 14
spreadsheets, 50
standard of living, 73
statistics, 21–23, 48, 96; "flattening the curve" strategy, 85–86
stereotype, 140
Stiglitz, Joseph E., 73
storage: Big Data, 89; data, 91; information, 7–8
storytelling, 26–27
subjectivity, 96
supply chain, 54, 60–61, 99; applications, 51; artificial intelligence (AI), 50–51; automobile manufacturing, 48–49; barcodes, 50; informed, 48; infrastructure, 48–49; just-in-time approach, 50; sensors, 49–50
surveillance, 63, 86, 104; capitalism, 105; disease, 86–87. *See also* privacy
swarm behavior, 158–59

system(s), 5, 13, 144–45, 148; adaptive, 19–20; computer, 38, 150–51; operating, 149; optimization, 151

tabulating cards, 83–84
taxonomy, 7
technological determinism, 32, 91, 95
technology, 4, 27–29, 71, 149; convergence, 54; platform, 60, 151, 153; weaponization, 119. *See also* artificial intelligence (AI); computer/computing; robotics
telecommunication, 14, 52–53
telegraphy, 6, 67–68
télématique, 81
television, 71
tendency, 10
terrorism, 9/11 attacks, 123–25
Tetlow, Philip, 4–5, 31–32
theory, 156; -based science, 105–7; conspiracy, 112, 116, 118; of evolution, 23
Toffler, Alvin, 81–83; *Future Shock*, 78–79, 102
training, 70; AI, 175; on-the-job, 8; librarian, 130
tribalism, 76–77, 141–42
Trump, Donald, 77, 111–12, 119, 122–23, 125, 128; tweets, 129
trust, 66
truth, 4, 17, 26–27, 87, 101, 113, 130; misrepresentation, 114–15
turnpike effect, 29

United States, 35–36; 9/11 attacks, 123–25; agricultural sector, 11–12; Electronic Signatures in Global and National Commerce Act (2000), 58; Federal Reserve System, 49; manufacturing, 40
university, 72; library, 70–71
urban legend, 123–24. *See also* conspiracy theory
U.S. Food and Drug Administration, 115

value(s), 10, 20, 73, 113, 131, 139, 157; Big Data, 90; data, 89
variety, Big Data, 89–90
veracity, Big Data, 90, 109
Voltaire, 1, 3, 32

Wal-Mart, 92
Wardle, Claire, 117
Washington, George, 116
Washington Post, 111
Watson: advertising, 172; marketing, 170–72
Watson, James, 3
Watson, Thomas J., 132
weaponized information, 118–19
web crawlers, 92
Weller, Toni, 77–78
"what if" analysis, 38–39, 48
Wiener, Norbert, 14, 178
Wikipedia, 111, 146
Williams, Deirdre, 129
Williams, Rosalind, 4; "rolling apocalypse", 20
Wired, 106
work: collaboration, 54–55; digital style, 61; digitization, 58–59; ethic, 157; from home, 52; knowledge, 72
workforce: agriculture, 11–12; AI impacts, 174; expansion of occupational groups, 35–37, 70; healthcare, 36–37; information-intensive, 36–37; management, 36; production, 36. *See also* labor
World Wide Web, 92
writing, 72–73

YouTube, 103

Zuboff, Shoshana, 108; *The Age of Surveillance Capitalism: The Fight for a Human Future at the New Frontier of Power*, 104–5; "You Are Now Remotely Controlled", 105
Zuckerberg, Mark, 127, 146

About the Author

James W. Cortada is senior research fellow at the Charles Babbage Institute at the University of Minnesota–Minneapolis. He holds a PhD in modern history and spent nearly four decades working at IBM in various sales, managerial, and research positions. He has spent nearly a half century at the center of developments concerning the world's engagement with computers, and he has written on the computer's contemporary uses in business publications, the history of computing, and the history of information. He has published more than a dozen books on the role of information in modern society, including *Fake News Nation: The Long History of Lies and Misinterpretations in America*, *Building Blocks of Society: History, Information Ecosystems, and Infrastructures*, and *Birth of Modern Facts: How the Information Revolution Transformed Academic Research, Governments, and Businesses*.

www.ingramcontent.com/pod-product-compliance
Ingram Content Group UK Ltd.
Pitfield, Milton Keynes, MK11 3LW, UK
UKHW022118060425
457132UK00008B/45